Lubrication and Dynamic Characteristics Analysis Technology

Bin Zhao · Xuan Ma · Xiujiang Shi · Xiqun Lu · Zhijun Shuai · Wanyou Li

Lubrication and Dynamic Characteristics Analysis Technology

For Journal-Thrust Coupled Bearing and Double-Layer Oil Film Floating Bearing in Marine Engine

Springer

Bin Zhao
Harbin Engineering University
Harbin, China

Xuan Ma
Harbin Engineering University
Harbin, China

Xiujiang Shi
Harbin Engineering University
Harbin, China

Xiqun Lu
Harbin Engineering University
Harbin, China

Zhijun Shuai
Harbin Engineering University
Harbin, China

Wanyou Li
Harbin Engineering University
Harbin, China

ISBN 978-981-95-2269-9 ISBN 978-981-95-2270-5 (eBook)
https://doi.org/10.1007/978-981-95-2270-5

This Springer imprint is published by the registered company Springer Nature Singapore Pte Ltd.
The registered company address is: 152 Beach Road, #21-01/04 Gateway East, Singapore 189721, Singapore

Preface

Lubrication performance is the key to the service performance of diesel engine sliding bearings, and lubrication failure is also a common fault during the operation of diesel engines. Designing and evaluating lubrication performance based on different demands is an important part of the research and development process. Continuously enhancing the ability to design and evaluate bearing lubrication in actual machine environments is the key to achieving independent development of bearings. The research on the lubrication performance of traditional diesel engine sliding bearings has a long history, but the research on the lubrication performance of diesel engine crankshaft radial-thrust bearings and fuel pump roller-floating bushing-roller pin structures is still relatively rare.

Under the support of national basic scientific research projects such as the National Natural Science Foundation of China, the author has carried out a large amount of research work on the lubrication characteristics and dynamic characteristics of diesel engine crankshaft radial-thrust bearings and fuel pump roller-floating bushing-roller pin structures. The book consists of seven chapters. Among them, Chap. 1 generally introduces the form and function of the crankshaft journal—thrust coupling bearing of diesel engines. Chapter 2 establishes a thermal elastic flow lubrication model for textured journal—thrust coupling bearings, revealing the influence laws of eccentricity ratio, thrust element film thickness and texture distribution on lubrication performance and dynamic characteristics. Chapter 3 establishes a transient thermal-elastic flow lubrication model for JTC bearings of Marine diesel engines under time-varying working conditions. Chapter 4 introduces the multi-objective particle swarm optimization algorithm to form a collaborative optimization design method for the dual-surface texture of radial-thrust bearing shells. Chapter 5 introduces the working characteristics of high-pressure fuel pumps and the lubrication methods of floating bushing structures. Chapter 6 establishes a steady-state solid-liquid multi-physics field coupling analysis method for floating bushings. Chapter 7 takes the transient operating conditions of high-pressure fuel pumps into account in the floating bushing lubrication model, forming a floating bushing lubrication analysis method that is more in line with the timing state. The publication of this book supplements the lubrication design theories and methods of two special forms of sliding bearings and has certain scientific significance and practical engineering value for relevant engineering and technical personnel. In addition, this book can also serve as a reference material for engineering and technical personnel engaged in design and research in fields such as machinery, manufacturing, and national defense weapons.

The research achievements involved in this book are derived from the basic scientific research projects led by the author, such as the National Natural Science Foundation of China project (Project No. 52271312) "Analysis and Intelligent Optimization of Transient Lubrication Characteristics of Radial-Thrust Composite Bearing Shells Coupled with Three-Dimensional Vibration of CrankShafts", which embodies the collective creativity and wisdom of the author's research team. Here, we express our deep gratitude to all the funding departments. The research achievements of the author's research team have provided support for the content of this book. Here, we express our gratitude to them. In addition, this book has referred to and cited many relevant domestic and foreign literatures, which have been included in the references of this book. We would like to express our gratitude to all of them here. Due to the author's limited ability and the fact that the content described in this book is still in the process of development, there may be omissions or deficiencies in the book. We sincerely invite all readers to offer their criticism and corrections.

Harbin, China
July 2025

Bin Zhao

Contents

Part I
Journal-Thrust Coupled Bearings for Diesel Engine Crankshaft

Chapter 1
Overview of Journal-Thrust Coupled Bearings for Diesel Engine Crankshafts

Keywords Marine diesel engine · Journal-thrust bearing · Coupled model · Three-dimensional temperature field

1.1 Introduction

The journal-thrust coupled (JTC) bearing in diesel engine (as shown in Fig. 1.1) operates under complex load and high working temperature. Under extreme working conditions, it supports the crankshaft and prevents its axial movement through the lubrication of journal and thrust parts. Unlike conventional journal bearing and thrust bearing, the JTC bearing contains both these two parts and there is a definite coupled relationship between the two parts. At present, there are few studies on this type of bearing, and the research on the continuity mechanism of the journal part and thrust part of the bearing is still unclear. This chapter introduces the form and function of the journal-thrust coupled bearing for diesel engine crankshaft.

1.2 Background and Meaning of the Research

With the rapid development of diesel engine technology towards higher power density, lower emissions, and higher reliability, the design and optimization of its core components have become key research topics in recent years. As a critical component that supports the crankshaft, the bearing plays a crucial role in the safety of both the crankshaft and the entire engine. During the development of new engine models, domestic production of bearings has become a crucial part of the process. However, domestic bearing manufacturers have mostly relied on imported patents, with insufficient research and development capabilities. Moreover, foreign countries

B. Zhao et al., *Lubrication and Dynamic Characteristics Analysis Technology*,
https://doi.org/10.1007/978-981-95-2270-5_1

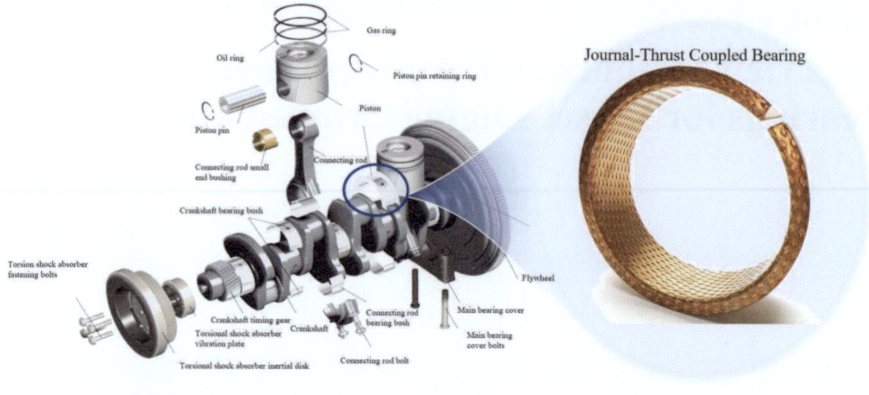

Fig. 1.1 Journal-thrust coupled bearing in diesel engine crankshaft system

have imposed technological restrictions on the core technology of bearing development, resulting in a significant performance gap between domestic and imported bearings. This has made the development of high-performance bearings a "critical bottleneck" in the domestic production of diesel engines. Lubrication performance is a key component of bearing service performance, and bearing lubrication failure is one of the most common faults that occur during diesel engine operation. Designing lubrication systems based on different needs to improve lubrication service performance is an essential part of bearing development. While there has been a long history of research into bearing lubrication, accurately and in real time assess the transient lubrication characteristics of bearings in the complex dynamic environment of diesel engines remains a challenging issue.

Journal-thrust coupled bearings, as a special type of bearing at the end of the crankshaft system in diesel engines, serve the dual purpose of supporting the crankshaft and preventing its axial movement. These bearings operate in extremely harsh environments, often under high temperatures, high loads, and high speeds. Under both transverse and axial loads, the journal and thrust parts of the bearing form two friction pairs with the crankshaft journal and thrust shoulder. As a result, this type of bearing is more prone to lubrication failure under the effects of longitudinal and transverse vibrations of the crankshaft. In recent years, during the development and service of domestically produced marine diesel engines, journal-thrust coupled bearings have experienced lubrication-related failures in different engine models multiple times. However, the lubrication mechanism of journal-thrust coupled bearings is still not fully understood. Therefore, establishing a coupled lubrication model suitable for journal-thrust coupled bearings has significant theoretical and engineering value.

The design of journal-thrust coupled bearings is relatively unique, as it includes both journal and thrust parts. Under normal operating conditions, the bearing is only subjected to radial loads, and can be considered a journal bearing. However, when axial movement is significant, the gap between the thrust part and the thrust shoulder becomes smaller, and the oil film begins to bear a load. At this point, the journal-thrust coupled bearing carries load in both directions. Due to the integrated design of the journal-thrust coupled bearing, the thrust part is lubricated by oil supplied through the journal part's end leakage, so it cannot be treated as a simple combination of a journal part and a thrust part. It is necessary to explore the coupled relationship between these two parts and develop a more accurate lubrication model.

The crankshaft, as the core transmission component in a diesel engine, is subjected to periodic pressure changes from the connecting rod during operation. Due to its three-dimensional spatial structure, the operating conditions and dynamic behavior are quite complex. This poses challenges for the transient lubrication analysis of journal-thrust coupled bearings. On the one hand, the radial movement of the bearing is directly influenced by radial cyclic loads; on the other hand, the thrust part is directly affected by axial displacement. Therefore, the transient lubrication performance under multi-directional dynamic loading conditions requires further investigation.

After accurately modeling the lubrication of the journal-thrust coupled bearing under actual operating conditions, targeted optimization of the bearing structure can be carried out with the aim of improving its tribological and dynamic performance. A large number of studies have shown that surface texturing can effectively improve lubrication conditions, reduce friction and wear, and is an effective means of improving the load-carrying capacity and service life of bearings. Considering the dual structure of journal-thrust coupled bearings, it is essential to investigate how to process appropriate textures on both parts to balance the load-carrying capacity of the thrust part with the friction-reducing ability of the journal part.

1.3 Introduction of Lubrication Performance Analysis for Conventional Sliding Bearings

The research on sliding bearing lubrication originated in the late nineteenth century, with the initial lubrication theory being based on fluid dynamic lubrication principles. In 1883, Petrov conducted experimental studies on sliding bearings and discovered that friction losses in bearings were closely related to the viscosity of the lubricant, particularly the viscous shear effects in the oil film that significantly influenced friction. This research laid the foundation for the development of lubrication theory. Meanwhile, Tower (1883, 1885) also observed in his experiments that the pressure inside bearings was not uniformly distributed and that the oil film generated higher pressures in certain areas. This phenomenon provided the theoretical basis for the fluid dynamic pressure lubrication analysis. Based on these findings, Reynolds (Reynolds 1885), in 1886, proposed the fundamental equation of fluid dynamic pressure lubrication—the Reynolds equation. By simplifying the equations of motion for viscous fluids and assuming constant viscosity, density, and temperature of the fluid, he derived an equation to describe fluid lubrication phenomena. The Reynolds equation revealed the intrinsic relationship between the oil film pressure distribution and oil film thickness, providing the theoretical foundation for subsequent research.

With the continuous development of computational methods and theoretical tools, the Reynolds equation became one of the core theories of fluid dynamic lubrication. In the early twentieth century, many scholars, building upon the Reynolds equation, further explored various issues related to bearing lubrication and conducted a series of experimental validations. Due to the limited computational tools available at the time, many scholars simplified the Reynolds equation for different bearing types and operating conditions to facilitate its solution. Sommerfeld (1914) and Ocvirk (1952) applied different simplifications for infinitely long bearings and infinitely short bearings, respectively, and obtained analytical solutions, providing theoretical support for the analysis of bearing lubrication performance.

In the 1950s, the advancement of computer technology and numerical methods brought a significant leap in lubrication theory research. In 1956, Pinkus was the first to use numerical methods to study the lubrication performance of circular and elliptical bearings, pioneering the use of numerical methods in bearing lubrication analysis. As computing power continued to improve, scholars began using numerical methods to simulate more complex lubrication problems, solving many issues that traditional analytical methods could not address. In particular, considering factors such as the viscosity-temperature and viscosity-pressure effects in fluid dynamic lubrication further refined lubrication theory. Many studies have shown that the viscosity of lubricating oil changes with temperature and pressure, and the thermal effects of the lubricant have a significant impact on oil film pressure distribution, oil film thickness, and bearing load capacity. As a result, more and more scholars began to incorporate factors such as oil film temperature variations, thermal deformation of the lubricant, and elastic deformation into lubrication performance analysis.

As research deepened, the impact of temperature effects on bearing lubrication performance received increasing attention. Particularly with the widespread application of bearings under high-speed and high-load conditions, thermal effects on lubrication performance gradually became an important research topic in lubrication theory. The study of thermal fluid dynamic lubrication (THD) theory began to move from traditional isothermal assumptions to methods that consider thermal effects. Kim et al. (1983), in their study of tilting pad thrust bearings, found that two-dimensional temperature field analysis could not meet practical requirements, and three-dimensional temperature field analysis was necessary to accurately reflect the temperature changes and lubrication performance of bearings under thermal loading. Subsequently, Tieu et al. (1975) used finite element methods to obtain results for three-dimensional temperature fields and two-dimensional oil film pressure distributions, showing that three-dimensional temperature field analysis is indispensable for bearing thermal analysis. In 1984, Lund and Hansen proposed an improved model, assuming that the viscosity and temperature of the oil film followed a fourth-order polynomial distribution along the oil film thickness direction. This assumption allowed for a more precise description of the thermal characteristics of the oil film in thermal fluid dynamic lubrication calculations. In the 1990s, Vijayaraghavan (1996) introduced the Lobatto point method, using Legendre polynomials to describe the viscosity and temperature distribution of the oil film. This method significantly improved the computational efficiency of thermal fluid dynamic lubrication and laid the foundation for more precise simulations.

With the increasing use of bearings in extreme operating conditions (such as high load and high speed), scholars began focusing on the effects of elastic deformation and thermal deformation of lubrication surfaces on lubrication performance. The theory of thermoelastic fluid dynamic lubrication (TEHD) emerged, considering both the elastic and thermal deformation of lubrication surfaces, as well as the thermal properties of viscosity and density of lubricating oil on lubrication performance. TEHD theory combines fluid dynamic lubrication with thermal and elastic deformation effects, enabling more accurate descriptions of bearing lubrication performance under high load and high-speed conditions. In 1985, Gethin et al. analyzed journal bearings with axial oil grooves and found that the thermal elastic deformation of the bearing had a significant impact on bearing load capacity and temperature distribution. Zhang et al. (2000) also experimentally verified that under heavy-load conditions, considering thermal deformation of the bearing shell resulted in a significant reduction in cavitation temperature, further proving the importance of thermal deformation on lubrication performance. As research in this field advanced, scholars began combining TEHD theory with different bearing designs and operating conditions for more detailed studies. For example, Fatu et al. (2006) introduced the concept of boundary layers to simplify the solution of TEHD equations, greatly improving the efficiency of the TEHD model and making it applicable to more complex bearing lubrication problems.

Although TEHD theory provides theoretical support for the stable operation of bearings, it still cannot effectively explain the changes in lubrication performance under transient operating conditions, especially when loads, speeds, or other conditions change rapidly. In these cases, traditional steady-state TEHD theory cannot accurately predict bearing failure. To address this, scholars proposed transient thermoelastic fluid dynamic lubrication (TTEHD) theory. TTEHD theory considers the time-dependent effects on bearing lubrication performance, especially during transient conditions such as bearing startup, acceleration, and deceleration. It captures the dynamic changes in oil film pressure, oil film thickness, and temperature field. In 1974, Rohde and Ezzat proposed a transient lubrication analysis method based on Reynolds and energy equations, finding that the thermal response time of the oil film was much shorter than that of the bearing pads, meaning that transient lubrication analysis must specifically consider the impact of temperature on lubrication state. As research into transient TEHD lubrication deepened, many scholars developed solution methods based on transient equation systems, widely applying them to different types of bearings (e.g., journal bearings, thrust bearings). Rodkiewicz and Venkateswarlu (1982) and Venkateswarlu and Rodkiewicz (1981) analyzed the transient characteristics during journal acceleration and found that bearing load capacity improved under transient conditions, though temperature effects still could not be ignored.

In addition, the effect of surface roughness on lubrication performance has consistently attracted widespread attention from scholars. Early fluid dynamic lubrication theories assumed that bearing surfaces were smooth and in full-film lubrication;

however, in reality, bearings often exhibit surface roughness, which can significantly affect the oil film thickness distribution. When the oil film thickness approaches or is smaller than the surface roughness, the lubrication state typically enters the mixed lubrication regime, where the interaction between surface roughness and oil film pressure has a significant impact on lubrication performance. In 1969, Christensen proposed a random model for surface roughness, providing a theoretical basis for subsequent studies on the effects of surface roughness on lubrication performance. Many scholars have since proposed lubrication models that consider surface roughness, such as the average flow model and micro-contact models, further improving the accuracy of lubrication performance predictions.

In conclusion, research into sliding bearing lubrication has evolved from the early fluid dynamic lubrication theory, with continuous improvement in bearing lubrication analysis methods. Scholars have gradually enriched lubrication theory by considering factors such as fluid dynamic lubrication, elastic/thermal deformation of bearings, oil film thermal effects, journal tilt, and solid contact (Zhou et al. 2021; Xie and Zhu 2021). As a result, lubrication theories such as fluid dynamic lubrication, elastic fluid dynamic lubrication, and thermal elastic fluid dynamic lubrication have been developed, culminating in the current model for transient mixed thermal-elastohydrodynamic lubrication of journal bearings and journal-thrust coupled bearings.

1.4 Introduction of Analysis for Journal-Thrust Coupled Bearings

Journal-thrust coupled bearings, as a special type of bearing, are widely used in mechanical equipment that operates under complex conditions, such as heavy machinery, ships, aircraft engines, automotive turbochargers, and computer hard disk drives. In these applications, the bearing not only needs to bear both axial and radial loads, but also must withstand high loads, high temperatures, and high-frequency vibrations. Therefore, its lubrication performance has a critical impact on the operational efficiency and service life of the equipment.

Journal-thrust coupled bearings typically consist of two parts: the journal part and the thrust part. The journal part primarily bears radial loads from the shaft, while the thrust part mainly bears axial loads. In practical bearing operation, the lubrication oil for the thrust part is typically supplied by the end leakage from the journal part, making the lubrication processes of the two parts interdependent. Since the operating environments and load conditions of the journal and thrust parts are different, their lubrication performance significantly influences each other. Therefore, to accurately predict the lubrication behavior of the entire bearing system, it is necessary to consider the coupled effects between the two parts.

Traditional bearing lubrication models often treat the journal and thrust parts as independent parts and analyze them separately. While this approach simplifies the calculations, it overlooks the physical coupled relationship between the two parts. Such simplification, especially under high-load and high-speed conditions, often leads to discrepancies in lubrication performance predictions. As a result, researchers have begun to explore coupled lubrication models for journal-thrust coupled bearings, achieving significant advancements.

In lubrication studies of journal-thrust coupled bearings, Jang and Yoon (2002) were the first to analyze the dynamic characteristics of journal-thrust coupled bearings under static loads and proposed a model that considers the combined effects of the journal and thrust parts. Later, Kim et al. (2011) introduced a stiffness and damping coefficient calculation method based on five degrees of freedom, providing a theoretical foundation for coupled lubrication analysis. These studies laid an important foundation for subsequent model improvements and research on coupled effects.

Building on this, Liu et al. (2021) further developed an elastic fluid dynamic lubrication (EHD) model under high environmental pressure, considering the effects of lubrication oil pressure and temperature, as well as the deformation and misalignment of the journal and thrust parts. This model more accurately describes the lubrication behavior of bearings under practical operating conditions. However, many existing models still have limitations and fail to fully consider the impact of bearing surface texture, temperature field distribution, and thermoelastic deformation on lubrication performance. Therefore, how to account for more real-world factors into the coupled model has become a hot topic in current research.

With the increasing depth of research into coupled effects, more scholars are beginning to consider the introduction of flow and pressure continuity conditions in the lubrication analysis of journal-thrust coupled bearings. These conditions can more accurately describe the interaction between the two parts. Wang et al. (2002, 2003) were the first to propose the coupled lubrication effects of journal-thrust coupled bearings and studied their lubrication performance based on fluid dynamic lubrication (HD) models, finding that the coupled effects between the two parts significantly affected oil film pressure, load-carrying capacity, and lubrication performance. To further improve model accuracy, Xiang et al. (2021a, b) incorporated elastic deformation considerations and developed an EHD model, further optimizing the precision of lubrication performance predictions. This method successfully revealed the coupled effects between the journal and thrust parts and provided a theoretical basis for the design and optimization of journal-thrust coupled bearings.

In addition to mechanical coupling, thermal effects also play a crucial role in the lubrication performance of journal-thrust coupled bearings. Under high-temperature conditions, the changes in lubricant viscosity and the temperature distribution of the oil film have a significant impact on lubrication performance. Stefani (2006) and Thomas et al. (2015) established a steady-state thermal fluid dynamic lubrication (THD) model, considering heat transfer effects, and initially explored the thermal coupled effects. They analyzed the temperature field using a simplified heat balance equation and found that temperature changes at the common boundary significantly affected the oil film temperature of both parts. Therefore, thermal coupled effects are an important factor influencing the lubrication performance of journal-thrust coupled bearings, especially under high-temperature working conditions. Effectively considering thermal elastic deformation and oil film temperature changes is a key direction for improving lubrication models.

Although significant progress has been made in the research on lubrication of journal-thrust coupled bearings, many challenges and unresolved issues remain, particularly in the field of diesel engines, requiring further in-depth exploration. Currently, most research focuses on theoretical modeling and static analysis. While lubrication performance under normal operating conditions can be accurately described, existing models often fail to fully consider the complex factors in the special operating environment of diesel engines. For example, the working environment of diesel engines typically involves extreme conditions such as high loads, high temperatures, and vibrations, which affect lubrication behavior, oil film stability, and load-carrying capacity. These impacts have not been comprehensively studied.

In particular, under high-temperature, high-load, and high-speed operating conditions, the lubrication performance of diesel engine bearings is significantly influenced by temperature fields and pressure fields. Thermal effects play a crucial role in oil film temperature changes, lubricant viscosity variations, and oil film thickness distribution. While some studies have considered thermal coupled effects, most models still simplify the temperature field calculations, failing to fully account for the coupling of thermal elastic deformation and lubricant property changes. This leads to imprecise predictions of lubrication performance under high-temperature diesel engine conditions. Therefore, coupled temperature fields, oil film thickness, and thermal elastic deformation remains a critical area for future research.

For journal-thrust coupled bearings in diesel engines, conducting in-depth studies on coupled lubrication mechanisms under extreme conditions such as high temperature, high load, and high vibration is crucial. Developing more accurate lubrication models and using multi-physics coupled analysis methods is expected to have a profound impact on the design, optimization, and fault diagnosis of diesel engine bearings. These studies not only provide theoretical support for bearing lubrication performance but also offer important references for predicting the operational reliability and service life of related mechanical equipment under real-world conditions.

1.5 Introduction of Research on Surface Texture Optimization

Since Hamilton et al. proposed in 1966 that surface texturing provides additional dynamic pressure lubrication effects, the mechanisms, modeling methods, and experimental studies of surface texturing have been hot research topics in the field of tribology. As mentioned in the background, the mechanism by which surface texturing improves lubrication performance of friction pairs differs under different lubrication conditions. The micro-pressurization effect is the most significant mechanism of surface texturing, which has attracted wide attention from scholars both domestically and internationally (Gropper et al.2016).

Literature Etsion et al. (1999), Ronen et al. (2001) and Etsion (2005) points out that local cavitation effects lead to asymmetric pressure distribution in the textured units, which generates additional fluid lift. Qiu and Khonsari (2009, 2011), as well as Zhang and Meng (2012), captured cavitation phenomena inside individual textured regions through high-speed cameras in their experiments. However, in some cases, the asymmetric pressure distribution in the texture may also have a negative effect on load-bearing capacity. Olver et al. (2006) and Fowell et al. (2007) noted that the textured surface only provides additional fluid lift when the supply oil pressure exceeds the cavitation pressure. This is because a higher supply pressure generates a pressure gradient between the lubricated surfaces, thereby ensuring proper lubrication oil supply.

Placing surface textures in the oil inlet region is another method to increase load-bearing capacity. As long as the supply pressure is higher than the cavitation pressure, textures in the oil inlet region will provide additional fluid lift. Tønder (1996, 2004) suggested introducing surface roughness or waviness in the inlet region, which can create a stepped structure similar to a Rayleigh step bearing, thereby significantly enhancing load-bearing capacity. The authors also found that surface roughness or waviness in the oil inlet region can effectively reduce oil leakage, allowing more lubrication medium to remain in the pressure-bearing region.

Another explanation for the disturbance in pressure distribution within the textured units is the fluid inertia effect, first proposed by Arghir et al. in 2003. Since simplified models based on Reynolds or Stokes equations do not consider fluid inertia effects, the authors solved the full Navier–Stokes equations for different macro-roughness units and found that fluid inertia effects provide additional fluid lift for fluids with higher Reynolds numbers. Later, Sahlin et al. (2005) used CFD to study fluid inertia effects in textured units and found that fluid inertia is the primary cause of dynamic pressure generation. Their study also indicated that the bearing capacity of textures reaches its maximum when the texture depth approaches the depth at which vortices occur. However, Dobrica and Fillon (2009) presented completely contrary results. They studied two-dimensional textured parallel sliders and pointed out that fluid inertia effects can have a negative impact on load-bearing capacity. Kraker et al. (2010) emphasized that the impact of fluid inertia effects on load-bearing capacity in textured units depends on the local flow conditions of the lubrication medium. Cupillard et al. (2008a, b, 2009, 2010) conducted systematic studies on textured sliders and bearings using CFD. In reference Strzelecki (2005), the authors noted that fluid inertia effects in textures do not necessarily have a positive impact on load-bearing capacity. In fact, only when the texture depth reaches a certain critical value does the load-bearing capacity increase.

Etsion's team has demonstrated the enormous potential of local texturing in mechanical components such as mechanical seals (Etsion and Halperin 2002), thrust bearings (Brizmer et al. 2003; Etsion et al. 2004), piston rings (Ryk et al. 2005; Ryk and Etsion 2006), and journal sliding bearings (Brizmer and Kligerman 2012).

The aforementioned research indicates that local texturing can improve tribological performance through a so-called collective effect, which is similar to Tønder (1996, 2004) findings. It is worth noting that local texturing, similar to a Rayleigh step bearing, its fluid dynamic pressure effect is determined only by its geometric shape. However, compared to traditional step bearings, local texturing has several advantages, such as collecting surface wear particles and storing lubricating oil.

It should be noted that the load-bearing mechanisms of the surface texturing mentioned above may occur simultaneously. Therefore, the influence of texturing on tribological performance depends on the structural parameters and operating conditions of the mechanical components. Engineers must adopt a "problem-specific analysis" approach when designing surface textures.

References

Arghir M, Roucou N, Helene M et al (2003) Theoretical analysis of the incompressible laminar flow in a macro-roughness cell. J Tribol 125(2):309–318

Brizmer V, Kligerman Y (2012) A laser surface textured journal bearing. J Tribol 134(3):031702

Brizmer V, Kligerman Y, Etsion I (2003) A laser surface textured parallel thrust bearing. Tribol Trans 46(3):397–403

Christensen H (1969) Stochastic models for hydrodynamic lubrication of rough surfaces. Proc Inst Mech Eng 184(1):1013–1026

Cupillard S, Cervantes M, Glavatskih S (2009) Pressure build-up mechanism in a textured inlet of a slider bearing. Res Report 33:33

Cupillard S, Glavatskih S, Cervantes MJ (2008a) Computational fluid dynamics analysis of a journal bearing with surface texturing. ARCHIVE Proc Inst Mech Eng Part J J Eng Tribol 222(2):97–107 [1994–1996 (vols 208–210)]

Cupillard S, Cervantes M, Glavatskih S (2008b) A CFD study of a finite textured journal bearing. In: IAHR symposium on hydraulic machinery & systems:27/10/2008–31/10/2008

Cupillard S, Glavatskih S, Cervantes MJ (2010) Inertia effects in textured hydrodynamic contacts. ARCHIVE Proc Inst Mech Eng Part J J Eng Tribol 224(8):751–756 [1994–1996 (vols 208-210)]

Dobrica MB, Fillon M (2009) About the validity of Reynolds equation and inertia effects in textured sliders of infinite width. Proc Inst Mech Eng Part J J Eng Tribol 223(1):69–78

Etsion I (2005) State of the art in laser surface texturing. J Trib 127(1):248–253

Etsion I, Halperin G (2002) A laser surface textured hydrostatic mechanical seal. ASLE Trans 45(3):430–434

Etsion I, Kligerman Y, Halperin G (1999) Analytical and experimental investigation of laser-textured mechanical seal faces. ASLE Trans 42(3):511–516

Etsion I, Halperin G, Brizmer V et al (2004) Experimental investigation of laser surface textured parallel thrust bearings. Tribol Lett 17(2):295–300

Fatu A, Hajjam M, Bonneau D (2006) A new model of thermoelastohydrodynamic lubrication in dynamically loaded journal bearings. J Tribol 128(1):85–95

Fowell MT, Olver AV, Gosman AD et al (2007) Entrainment and inlet suction: two mechanisms of hydrodynamic lubrication in textured bearings. J Tribol 129(2):336–347

Gethin DT (1985) An investigation into plain journal bearing behaviour including thermo-elastic deformation of the bush. Proc Inst Mech Eng C J Mech Eng Sci 199(3):215–223

Gropper D, Ling W, Harvey TJ (2016) Hydrodynamic lubrication of textured surfaces: a review of modeling techniques and key findings. Tribol Int 94:509–529

Hamilton DB, Walowit JA, Allen CM (1966) A theory of lubrication by micro-irregularities. J Basic Eng 88(1):177–185

Jang GH, Yoon JW (2002) Dynamic characteristics of a coupled journal and thrust hydrodynamic bearing in a HDD spindle system due to its groove location. Microsyst Technol 8:261–270

Kim KW, Tanaka M, Hori Y (1983) A three-dimensional analysis of thermohydrodynamic performance of sector-shaped, tilting-pad thrust bearings. J Lubr Technol 105(3):406–412

Kim H, Jang G, Lee S (2011) Complete determination of the dynamic coefficients of coupled journal and thrust bearings considering five degrees of freedom for a general rotor-bearing system. Microsyst Technol 17:749–759

Kraker AD, Ostayen R, Rixen DJ (2010) Development of a texture averaged Reynolds equation. Tribol Int 43(11):2100–2109

Liu R, Jing L, Meng X et al (2021) Mixed elastohydrodynamic analysis of a coupled journal-thrust bearing system in a rotary compressor under high ambient pressure. Tribol Int 159:106943

Lund JW, Tonnesen J (1984) An approximate analysis of the temperature conditions in a journal bearing. Part II Appl J Tribol 106(2):228–236

Ocvirk FW (1952) Short-bearing approximation for full journal bearings. Tech Rep Arch Image Lib

Olver AV, Fowell MT, Spikes HA, et al (2006) 'Inlet suction', a load support mechanism in non-convergent, pocketed, hydrodynamic bearings. ARCHIVE Proc Inst Mech Eng Part J J Eng Tribol 220(2):105–108 [1994–1996 (vols 208–210)]

Petrov NP (1883) Friction in machines and the effect of the lubricant. Inzh Zh St-Peterb

Pinkus O (1956) Analysis of elliptical bearings. Trans Am Soc Mech Eng 78(5):965–972

Qiu Y, Khonsari MM (2009) On the prediction of cavitation in dimples using a mass-conservative algorithm. J Tribol 131(4):041702

Qiu Y, Khonsari MM (2011) Experimental investigation of tribological performance of laser textured stainless steel rings. Tribol Int 44(5):635–644

Reynolds O (1885) On the theory of lubrication and its application to Mr. Beauchamp Tower's experiments, including an experimental determination of the viscosity of olive oil. Philos Trans R Soc 1:157

Rodkiewicz CM, Venkateswarlu K (1982) Thrust bearing in time-dependent turbulent flow. ASLE Trans 25(1):101–107

Rohde SM, Ezzat HA (1974) A study of thermohydrodynamic squeeze films. J Lubr Technol 96(2):198–205

Ronen A, Etsion I, Kligerman Y (2001) Friction-reducing surface-texturing in reciprocating automotive components. Tribol Trans 44(3):359–366

Ryk G, Etsion I (2006) Testing piston rings with partial laser surface texturing for friction reduction. Wear 261(7–8):792–796

Ryk G, Kligerman Y, Etsion I et al (2005) Experimental investigation of partial laser surface texturing for piston-ring friction reduction. Tribol Trans 48(4):583–588

Sahlin F, Glavatskih SB, Almqvist T et al (2005) 2D CFD-analysis of micro-patterned surfaces in hydrodynamic lubrication. J Tribol 127(1):96–102

Sommerfeld A (1914) Zur Theorie der Schmiermittelreibung. Electr Eng 3(1):1–5

Stefani F (2006) THD analysis of a coupled journal and tilting pad thrust bearing system. In: International conference on tribology. Parma, pp 20–22

Strzelecki S (2005) Operating characteristics of heavy loaded cylindrical journal bearing with variable axial profile. Mater Res 8(4):481–486

Tieu AK (1975) A numerical simulation of finite-width thrust bearings, taking into account viscosity variation with temperature and pressure. J Mech Eng Sci 17(1):1–10

Tower B (1883) First report on friction experiments. Proc Inst Mech Eng 34(1):632–659

Tower B (1885) Research committee on friction: second report on friction experiments. Proc Inst Mech Eng 36(1):58–70

Tϕnder K (1996) Dynamics of rough slider bearings: effects of one-sided roughness/waviness. Tribol Int 29(2):117–122

Tϕnder K (2004) Hydrodynamic effects of tailored inlet roughnesses: extended theory. Tribol Int 37(2):137–142
Venkateswarlu K, Rodkiewicz CM (1981) Thrust bearing characteristics when the slider is approaching terminal speed. Wear 4(3):341–350
Vijayaraghavan D (1996) An efficient numerical procedure for thermohydrodynamic analysis of cavitating bearings. J Tribol 118(3):555–563
Wang Y, Wang QJ, Lin C (2002) Mixed lubrication of coupled journal-thrust bearing systems. Comput Model Eng Sci 3(4):517–530
Wang Y, Wang Q, Lin C (2003) Mixed lubrication of coupled journal-thrustbearing systems including mass conserving cavitation. J Tribol 125:747–755
Xiang G, Han Y, Wang J et al (2021a) Influence of axial microvibration on the transient hydrodynamic lubrication performance of misaligned journal–thrust microgrooved coupled bearings under water lubrication. Tribol Trans 64(4):579–592
Xiang G, Wang J, Zhou C et al (2021b) A tribo-dynamic model of coupled journalthrust water-lubricated bearings under propeller disturbance. Tribol Int 160:107008
Xie Z, Zhu W (2021) Theoretical and experimental exploration on the micro asperity contact load ratios and lubrication regimes transition for water-lubricated stern tube bearing. Tribol Int 164:107105
Zhang J, Meng Y (2012) Direct observation of cavitation phenomenon and hydrodynamic lubrication analysis of textured surfaces. Tribol Lett 46(2):147–158
Zhang C, Yi Z, Zhang Z (2000) THD analysis of high speed heavily loaded journal bearings including thermal deformation, mass conserving cavitation, and turbulent effects. J Tribol 122(3):597–602
Zhou G, Qiao J, Pu W et al (2021) Analysis of mixed lubrication performance of water-lubricated rubber tilting pad journal bearing. Tribol Int 169:107423

Chapter 2
Steady-State Thermal Elastic Hydrodynamic Lubrication and Dynamic Characteristics of Journal-Thrust Coupled Bearings

Abstract The journal-thrust coupled bearing in diesel engine operates under complex load and high working temperature. Under extreme working conditions, it supports the crankshaft and prevents its axial movement through the lubrication of journal and thrust parts. This chapter is mainly aimed to establish the thermal elastohydrodynamic (TEHD) lubrication analysis model for textured journal-thrust coupled bearing, which considered the thermal and pressure coupled effects in the lubrication process by proposing the flow, pressure and thermal continuity conditions between the journal and thrust parts. A specific coupled bearing testing equipment was set up and the experimental results validated the correctness of the TEHD lubrication model. Necessity of considering thermal and pressure coupled effect was confirmed through comparison of different models. Apart from that, parametric analyses were conducted to explore the influences of thermal and pressure coupled effects on lubrication performances and dynamic characteristics under different eccentricity ratio of journal part, geometry film thickness of thrust part and surface topography (mainly realized by texture distribution of thrust part).

Keywords Coupled bearing · Thermal coupled effect · Lubrication performances · Dynamic characteristics

2.1 Introduction

The lubrication performances of journal-thrust coupled bearings (Fig. 2.1) differ from those of conventional bearings (thrust/journal), while the lubrication analysis methods of conventional bearings are still the basis for exploring their lubrication characteristics. So far, researches about the lubrication characteristics of conventional bearings have been relatively comprehensive. On the basis of hydrodynamic (HD) lubrication theory, which has been recognized as the foundation of bearing lubrication, researchers have systematically considered the coupled effects of fluid (fluid lubrication), solid (rough contact (Cai et al. 2022a, b, c; Zhang et al. 2022) and thermoelastic deformation (Wang et al. 2002; Xie et al. 2022; Xie et al. 2023a, b)

B. Zhao et al., *Lubrication and Dynamic Characteristics Analysis Technology*,
https://doi.org/10.1007/978-981-95-2270-5_2

and heat (heat convection and heat conduction (Xie et al. 2021; Xu et al. 2022), and have successively developed elastohydrodynamic (EHD) (Liu et al. 2019; Liu et al. 2020a, b) lubrication and thermal elastohydrodynamic (TEHD) (Xiang et al. 2022; Liu et al. 2020a, b) lubrication analysis model. In general, the lubrication models of conventional bearing in diesel engine have been formed systematically.

Similarly, a lot of work has been done on the calculation and solution of dynamic characteristics of the conventional bearings, especially stiffness and damping. Since Klit and Lund (Klit et al. 1986) first proposed the linear perturbation method of Reynolds equation in 1986, relevant researchers began modifying the Reynolds equation to solve the dynamic characteristic parameters. Majumdar et al. (2016) analyzed stiffness and damping for journal bearing and proposed to solve the control equation with the first-order perturbation method and calculated. Pai et al. (2008) studied the stability of journal bearings with different number of grooves through nonlinear time transient analysis. Wang et al. (2018) improved the pressure perturbation method for thrust bearing and calculated stiffness and damping by considering the viscosity temperature effect of lubricant and the elastic deformation of bearing pad. Furthermore, many researchers have also studied the dynamic characteristics based on the commercial computational fluid dynamics (CFD) program (Guo et al. 2005). On the basis of extensive research, the lubrication and dynamic characteristic model with relatively complete consideration has been formed for conventional journal bearings and thrust bearings.

On the basis of research about conventional bearings, researchers also explored the lubrication performances and dynamic characteristics of journal-thrust coupled bearings. As can be found from the structure, the thrust part is lubricated by the oil supplied from the end of the journal part, so coupled effect of these two parts apparently exists. Researchers have studied the journal-thrust coupled bearings in computer hard disk drives (Jang et al. 2002), marine propellers (Sun et al. 2020), rotary compressors (Liu et al. 2021), ship stern shaft systems (Xiang et al. 2021), automobile turbochargers (Tian et al. 2012; Peixoto et al. 2019; Tian et al. 2013)

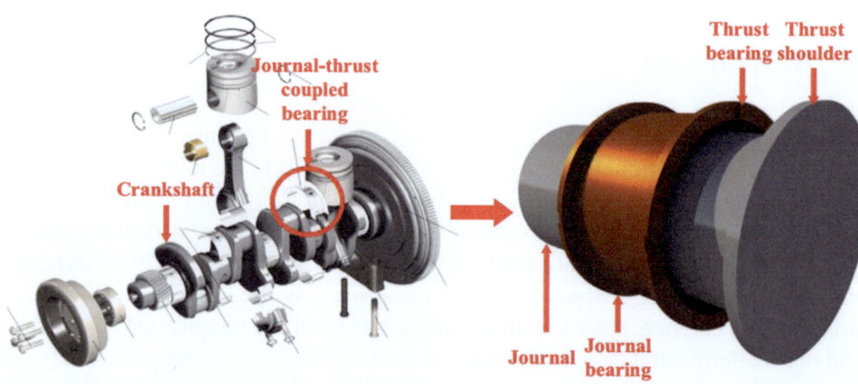

Fig. 2.1. Location and structure diagram of journal-thrust coupled bearing

and magnetic suspension turbomachinery (Han et al. 2016) so far. Jang et al. (2006) conducted a preliminary study on the static characteristics of the coupled bearing, and found the importance of common boundary conditions in the analysis of hydrodynamic performance. Later, Wang et al. (2003) explored the lubrication mechanism of the journal-thrust coupled bearing in the application of heavy machinery such as drilling rigs. In the way of unifying the lubrication calculation fields of journal part and thrust part, they proposed the flow and pressure continuity conditions of the journal-thrust common boundary and further evaluated the lubrication characteristics of the coupled bearing. Based on this theory, Xiang et al. (2019) took elastic deformation into consideration and established the EHD lubrication model, which further improved the precision of the model. They also made a preliminary exploration of the dynamic characteristics with consideration of coupled effect in the follow-up study (Xiang et al. 2021). The analysis results showed that the pressure coupled effect would significantly change pressure distribution, bearing capacity, stiffness and damping of both parts under EHD lubrication analysis. Furthermore, Stefani et al. (2006), (Hagemann et al. 2015) established a steady-state thermal hydrodynamic (THD) lubrication model for journal-thrust coupled bearing considering heat transfer. Without considering the thermoelastic deformation, they preliminarily explored the thermal coupled effect by using the simplified thermal balance equation. The results showed that the common boundary could greatly affect oil film temperature of each part.

To sum up, for the journal-thrust coupled bearing, TEHD analysis model considering the thermal and pressure coupled effects has not been formed at present. However, due to the severe and complex operating conditions for journal-thrust coupled bearing in diesel engine, the lubricant temperature is relatively high during the working process, thus leading the thermal effect and bearing deformation cannot be ignored. In order to better analyze the lubrication mechanism of journal-thrust coupled bearing under high temperature and heavy load conditions, the TEHD model is necessary to be establish. Furthermore, stiffness and damping calculation models also need to be studied for further crankshaft vibration analysis.

This chapter mainly focused on the journal-thrust coupled bearing of a specific marine diesel engine, which has texture on the thrust part to increase the axial bearing capacity. This chapter proposed flow, pressure and thermal continuity conditions of the journal-thrust common boundary and established the TEHD lubrication analysis model for journal-thrust coupled bearings. According to the model, this chapter explored the changing rules of lubrication characteristic parameters and dynamic characteristic parameters under different working conditions and different texture distribution, mainly including bearing capacity, oil film temperature, stiffness and damping, so as to provide theoretical support for accurately obtaining the operation condition of journal-thrust coupled bearing. It is worth noting that since the load on journal-thrust coupled bearings in actual diesel engines is dynamic, transient effects should be considered in the lubrication process. However, this chapter mainly focused on the coupled mechanism of the journal and thrust parts, and certain simplification has been made in the bearing load that the research is carried out under the condition of steady load.

2.2 Steady-State Lubrication and Dynamic Characteristics Analysis Model Considering Journal-Thrust Coupled Lubrication Effect

2.2.1 Film Thickness Model

The schematic diagram of the thrust part analyzed in this chapter is shown in Fig. 2.2a, and each pad surface of the thrust part has a certain inclination. The thrust part film thickness can be represented by Fig. 2.2b, and equation is given by:

$$h_{\mathrm{thrust}} = h_p + r\theta_{\mathrm{p}} \sin(\theta_0 - \theta) + \delta_{\mathrm{Ethrust}} + \delta_{\mathrm{Tthrust}} + h_{\mathrm{texture}} \tag{2.1}$$

where, h_p represents the thrust part geometry film thickness, which represents the minimum geometry distance from the thrust pad to the thrust shoulder; θ_{p} represents the inclination angle of thrust pad; θ_0 represents the circumferential angle of each pad; $\delta_{\mathrm{Ethrus}t}$ represents elastic deformation of thrust bearing; $\delta_{\mathrm{Tthrust}}$ represents thermal deformation of thrust bearing; h_{texture} represents the depth of texture. The texture profile studied in this chapter was rectangular and the details can be obtained from Fig. 2.2, so h_{texure} is constant. Since only one of the thrust bearing pads at two ends is subject to hydrodynamic pressure during the actual operation, this chapter only considered the lubrication of the thrust bearing on one side.

The film thickness of the journal part can be represented by Fig. 2.2c, and the equation is given by:

$$h_{\mathrm{journal}} = c(1 + \varepsilon \cos(\theta - \varphi)) + \delta_{\mathrm{Ejournal}} + \delta_{\mathrm{Tjournal}} \tag{2.2}$$

where, c represents the radial clearance; ε represents the eccentricity ratio, which can be defined as $\varepsilon = \sqrt{e_x^2 + e_y^2}/c$; φ represents the attitude angle of journal part; $\delta_{\mathrm{Ejournal}}$

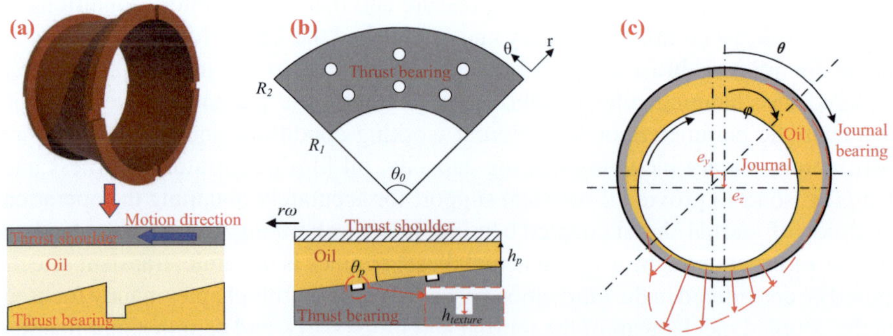

Fig. 2.2 Diagram of film thickness of the journal-thrust coupled bearing: **a** Schematic diagram of thrust part; **b** Film thickness of thrust part; **c** Film thickness of journal part

represents the elastic deformation of journal bearing; δ_{Tjournal} represents the thermal deformation of journal bearing. This chapter calculated the elastic deformation and thermal deformation of each part by using the influence coefficient method (Shi et al. 1998a; b).

2.2.2 Average Reynolds Equation Model

This chapter used the average Reynolds equation established by Patir and Cheng (Patir et al. 1978) to solve the oil film pressure. The Reynolds equation for the thrust part is as follows:

$$\frac{\partial}{r\partial\theta}\left(\phi_\theta\frac{h_{\text{thrust}}^3}{\eta}\frac{\partial p_T}{\partial\theta}\right)+\frac{\partial}{\partial r}\left(\phi_r\frac{rh_{\text{thrust}}^3}{\eta}\frac{\partial p_T}{\partial r}\right)=6\omega r\left(\phi_c\frac{\partial h_{\text{thrust}}}{\partial\theta}+\sigma\frac{\partial\phi_s}{\partial\theta}\right) \tag{2.3}$$

where, ϕ_θ, ϕ_r, ϕ_c and ϕ_s represent circumferential pressure-flow factor, radial pressure-flow factor, contact factor and shear-flow factor correspondingly; h_{thrust} represents thrust part film thickness calculated above; P_T represents the hydrodynamic pressure of thrust part; ω represents the relative speed of journal bearing. η represents the lubricant viscosity, which was calculated by Reynolds viscosity temperature equation $\eta=\eta_0\exp(\beta(T_{\text{actual}}-T_0))$ (Tucker et al. 1995), where η_0 represents the lubricant viscosity at reference temperature, β represents the viscosity temperature coefficient, T_{actual} represents the actual temperature, T_0 represents the reference temperature.

Furthermore, the Reynolds equation for the journal part is given by:

$$\frac{\partial}{R_0^2\partial\theta}\left(\phi_\theta\frac{h_{\text{journal}}^3}{\eta}\frac{\partial P_J}{\partial\theta}\right)+\frac{\partial}{\partial z}\left(\phi_z\frac{h_{\text{journal}}^3}{\eta}\frac{\partial P_J}{\partial z}\right)=\frac{6U_J}{R_0}\left(\phi_c\frac{\partial h_{\text{journal}}}{\partial\theta}+\sigma\frac{\partial\varphi_s}{\partial\theta}\right) \tag{2.4}$$

where, ϕ_z represents the axial pressure-flow factors; $h_{journal}$ represents the journal film thickness; P_J represents the journal hydrodynamic pressure; U_J represents the linear speed of journal part; R_0 represents the journal bearing radius.

2.2.3 Contact Model

The rough-surface contact model established by Lee and Ren (1996) was used in this chapter, which can be expressed as:

$$\begin{cases} \dfrac{\overline{h}(\gamma, H_Y, \overline{P})}{\sigma} = \exp\left\{\sum_{i=0}^{4} (\overrightarrow{\gamma}_G^T[G_i]H_Y)(\overline{P})^i\right\}, \overline{P} < H_Y \\ \dfrac{\overline{h}(\gamma, H_Y, \overline{P})}{\sigma} = 0.0,\ \overline{P} \geq H_Y \end{cases} \tag{2.5}$$

where, γG and [Gi] are parametric matrices given by the reference (Lee et al. 1996).

2.2.4 Thermal Model

The oil film temperature of each part can be further calculated through the energy equation. For thrust part, the circumferential and radial heat conduction is far less than the axial heat conduction, and also far less than the heat convection in their respective directions. Therefore, circumferential and radial heat conduction can be ignored and only axial heat conduction can be considered. At the same time, regardless of the axial pressure change, the assumption is that except for the velocity gradient of U_r and U_θ, other velocity gradients are ignored. The energy equation of thrust part (Jiang et al. 2011) is expressed as follows:

$$\rho C_p\left(U_r\frac{\partial T_T}{\partial r} + \frac{U_\theta}{r}\frac{\partial T_T}{\partial \theta}\right) = \alpha T\left(U_r\frac{\partial p_T}{\partial r} + U_\theta\frac{\partial p_T}{r\partial \theta}\right) + \lambda\frac{\partial^2 T_T}{\partial z^2} + \mu\left[\left(\frac{\partial U_r}{\partial z}\right)^2 + \left(\frac{\partial U_\theta}{\partial z}\right)^2\right] \tag{2.6}$$

where, C_p is the isopiestic specific heat; α is the thermal expansivity of lubricant; λ is the thermal conductivity of lubricant; U_r and U_θ are the radial speed and circumferential speed of the lubricant respectively. Compared with the heat convection and axial (z) heat conduction, the circumferential (θ) and radial (r) heat conduction of the thrust bearing is so small that they can be ignored in the equation.

For journal part, the circumferential and axial heat conduction is far less than the radial heat conduction, and also far less than the heat convection in their respective directions. Therefore, circumferential and axial heat conduction can be ignored and only radial heat conduction can be considered. At the same time, regardless of the radial pressure change, and the assumption is that except for the velocity gradient of v_J and w_J, other velocity gradients are ignored. The energy equation of journal part (Shi et al. 1998a; b) is expressed as:

$$\rho C_p\left(\frac{v_J}{r}\frac{\partial T_J}{\partial \theta} + w_J\frac{\partial T_J}{\partial z}\right) = \alpha T\left(w_J\frac{\partial p_J}{\partial z} + v_J\frac{\partial p_J}{r\partial \theta}\right) + \lambda\frac{\partial^2 T_J}{\partial r^2} + \mu\left[\left(\frac{\partial u_J}{\partial z}\right)^2 + \left(\frac{\partial v_J}{\partial z}\right)^2\right] \tag{2.7}$$

where u_J, v_J and w_J represent the radial velocity, circumferential velocity and axial velocity of the journal part respectively. Compared with the heat convection and radial (r) heat conduction, the circumferential (θ) and axial (z) heat conduction of the journal bearing is so small that they can be ignored in the equation.

Assuming that the bearing material is isotropic and the material properties are constant, so the equation of heat conduction is as follows, for thrust bearing:

$$\frac{\partial^2 T_T}{\partial r^2} + \frac{1}{r}\frac{\partial T_T}{\partial r} + \frac{1}{r^2}\frac{\partial^2 T_T}{\partial \theta^2} + \frac{\partial^2 T_T}{\partial z^2} = 0 \tag{2.8}$$

Similarly, for journal bearing:

$$\frac{\partial^2 T_J}{\partial x^2} + \frac{\partial^2 T_J}{\partial y^2} + \frac{\partial^2 T_J}{\partial z^2} = 0 \tag{2.9}$$

The temperature boundary conditions include:

(1) Oil inlet temperature boundary

$$T|_{\theta=0} = T|_{\theta=2\pi} = T_{in} = \frac{Q_s}{Q_s + Q_r}T_s + \frac{Q_r}{Q_s + Q_r}T_r \tag{2.10}$$

Q_s and Q_r are oil supply flow and return flow respectively, which are calculated by calculating the flow through the oil inlet control grid in both directions; T_s and T_r are oil supply temperature and return temperature respectively.

(2) Bearing-lubricant boundary

$$T_L = T_B,\ k_{rL}\partial T_L/\partial r = k_{rB}\partial T_B/\partial r \tag{2.11}$$

The subscript L is the lubricant parameter and the subscript B is the bearing parameter.

(3) Journal-lubricant boundary

Assuming that the inside of the journal is isothermal and isotropic:

$$k_{rJ}\partial T_J/\partial r = 1/2\pi k_{rL}\int_0^{2\pi} \partial T_L/\partial r d\theta \tag{2.12}$$

(4) External boundary conditions

The outside of the bearing is connected to the air, and there are thermal convection boundary conditions:

$$k\partial T/\partial n = -h_h(T - T_\infty) \tag{2.13}$$

where, n is the unit vector of convection direction, h_h is the convection heat transfer coefficient, and is the ambient temperature.

This chapter adopts the lubricant supply mode of supplying from the axial side of the bearing, so both the bearing-lubricant interface and bearing-environment interface should have continuous temperature and heat flux. The boundary conditions of the cavitation area are:

$$\frac{1}{\zeta} = \frac{1-\psi}{\zeta_{\text{gas}}} + \frac{\psi}{\zeta_{\text{lubricant}}} \tag{2.14}$$

where, ψ represents the rupture-position oil film thickness which is defined as: $\psi = \frac{h_{\min}}{h_{\text{begin}}}$, $h_{\min}$ represents the minimum film thickness in rupture-zone, h_{begin} represents film thickness at the starting position of rupture-zone; ζ represents thermal conductivity, specific heat and viscosity of different materials.

2.2.5 *Continuity Condition Model*

Due to the special structure of journal-thrust coupled bearing, the radial flow in the thrust part is linked with the axial flow in the journal part, which makes it necessary to consider the continuity conditions between the two parts. The continuity conditions diagram of journal-thrust coupled bearings is shown in Fig. 2.3. According to the research of Wang et al. (2003), continuity of circumferential flow is naturally satisfied, so ensuring continuity of radial flow in thrust part and axial flow in journal part is necessary. The specific expression is as follows:

$$\left(\frac{h_{\text{journal}}^3}{12\eta}\frac{\partial P_J}{\partial z}\right)_{j,m} = -\left(\frac{h_{\text{thrust}}^3}{12\eta}\frac{\partial P_T}{\partial r}\right)_{j,m} \tag{2.15}$$

where h and P represent the film thickness and hydrodynamic pressure near common boundary respectively, and the subscript represents the corresponding position.

On the basis of the research above, this chapter considers the thermal coupled effect and takes the continuity of heat flow into consideration, which ensure the continuity of radial heat flow of thrust part and axial heat flow of journal part. The axial heat flow of the journal part (Q_J) includes the axial heat convection and axial heat conduction, and the specific expression is as follows:

$$Q_J = \rho C_p w_J \frac{\partial T_J}{\partial z} + \lambda \frac{\partial^2 T_J}{\partial z^2} \tag{2.16}$$

Similarly, the radial heat flow of thrust part (Q_T) is expressed as:

$$Q_T = \rho C_p U_r \frac{\partial T_T}{\partial r} + \lambda \frac{\partial^2 T_T}{\partial r^2} \tag{2.17}$$

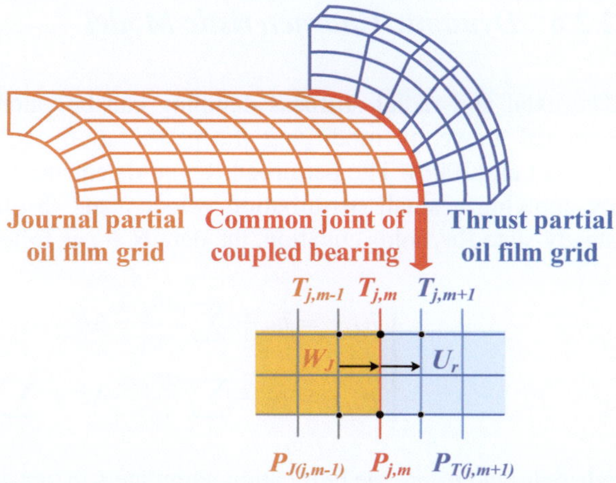

Fig. 2.3 Schematic diagram of continuity condition

The heat continuity condition between the journal part and the thrust part can be established by combining these two expressions. It should be noted that during the heat calculation, the axial heat conduction of the journal part and the radial heat conduction of the thrust part are far less than their heat convection, so the heat flow continuity formula can be simplified by ignoring the heat conduction term. In that case, the axial heat convection of the journal part needs to be equal to the radial heat convection of the thrust part. The final expression is as follows:

$$\left(w_J \frac{\partial T_J}{\partial z}\right)_{j,m} = -\left(U_r \frac{\partial T_T}{\partial r}\right)_{j,m} \tag{2.18}$$

The velocity of each part can be calculated using:

$$\begin{cases} w_J = \dfrac{h^3}{h_T^3} \phi_z \dfrac{1}{2\eta} \dfrac{\partial p_J}{\partial z} \left(d^2 - dh_T\right) \\ U_r = \dfrac{\partial p_T}{\partial r} \left(\displaystyle\int_0^z \frac{z\mathrm{d}z}{\eta} - \frac{\int_0^{h_{\text{thrust}}} \frac{\mathrm{d}z}{\eta}}{\int_0^{h_{\text{thrust}}} \frac{z\mathrm{d}z}{\eta}} \int_0^z \frac{\mathrm{d}z}{\eta} \right) \end{cases} \tag{2.19}$$

where $d = R_b\text{-}r$ represents distance from bearing surface to selected point.

By solving the Eq. (2.18) above with the finite difference method, the temperature at the common boundary can be obtained, as shown in the following equation:

$$T_{j,m} = \frac{U_r \Delta z T_{j,m+1} + w_J \Delta r T_{j,m-1}}{w_J \Delta r + U_r \Delta z} \tag{2.20}$$

where $T_{j,m}$ is the oil temperature in the common boundary area.

2.2.6 Dynamic Characteristic Model

The axial and radial stiffness damping can be calculated based on the pressure obtained above by modifying the perturbation equation method established by Jiang et al. (2006). The perturbed Reynolds equation in three directions is attained by applying a small perturbation velocity and displacement on the basis of the steady-state Reynolds equation, the details are as follows:

$$\begin{cases} h_{J/T} = h_0 + \sum \dfrac{\partial h_{J/T}}{\partial \zeta} \Delta \xi \\ p_{J/T} = p_0 + \sum \dfrac{\partial p_{J/T}}{\partial \zeta} \Delta \xi + \sum \dfrac{\partial p_{J/T}}{\partial \dot{\zeta}} \Delta \dot{\xi} \end{cases} \tag{2.21}$$

where h_0 and p_0 are the lubrication parameters in quasi equilibrium, while the others are perturbation terms. By substituting Eq. (2.21) into the film thickness and average Reynolds equation written above, the final equations are as follows:

$$\frac{\partial}{R\partial\theta}\left(\phi_\theta \frac{h_{\text{journal}}^3}{12\mu} \frac{\partial p_{J_\zeta}}{R\partial\theta}\right) + \frac{\partial}{\partial z}\left(\phi_z \frac{h_{\text{journal}}^3}{12\mu} \frac{\partial p_{J_\zeta}}{\partial z}\right)$$
$$= \begin{cases} -\frac{\partial}{R\partial\theta}\left(\phi_\theta \frac{h_{\text{journal}}^2}{4\mu} \frac{\partial p_{J_\zeta}}{R\partial\theta} \sin\theta\right) - \frac{\partial}{\partial z}\left(\phi_z \frac{h_{\text{journal}}^2}{4\mu} \frac{\partial p_{J_\zeta}}{\partial z} \sin\theta\right) + \frac{\omega R}{2} \cos\theta & : J_\zeta = x \\ -\frac{\partial}{R\partial\theta}\left(\phi_\theta \frac{h_{\text{journal}}^2}{4\mu} \frac{\partial p_{J_\zeta}}{R\partial\theta} \cos\theta\right) - \frac{\partial}{\partial z}\left(\phi_z \frac{h_{\text{journal}}^2}{4\mu} \frac{\partial p_{J_\zeta}}{\partial z} \cos\theta\right) + \frac{\omega R}{2} \sin\theta & : J_\zeta = y \\ 0 & : J_\zeta = z \\ \cos\theta & : J_\zeta = \dot{x} \\ \sin\theta & : J_\zeta = \dot{y} \\ 0 & : J_\zeta = \dot{z} \end{cases} \tag{2.22}$$

$$\frac{\partial}{r\partial\theta}\left(\phi_\theta \frac{h_{\text{thrust}}^3}{12\mu} \frac{\partial p_{T_\zeta}}{r\partial\theta}\right) + \frac{\partial}{\partial r}\left(\phi_r \frac{h_{\text{thrust}}^3}{12\mu} \frac{\partial p_{T_\zeta}}{\partial r}\right)$$
$$= \begin{cases} 0 & : T_\zeta = x \\ 0 & : T_\zeta = y \\ \frac{\partial}{r\partial r}\left(\frac{rh_{\text{thrust}}^2}{4\mu} \frac{\partial p_0}{\partial r}\right) + \frac{\partial}{r\partial\theta}\left(\frac{h_{\text{thrust}}^2}{4\mu} \frac{\partial p_0}{r\partial\theta}\right) & : T_\zeta = z \\ 0 & : T_\zeta = \dot{x} \\ 0 & : \xi_T = \dot{y} \\ 0 & : \xi_T = \dot{z} \end{cases} \tag{2.23}$$

where, J_ζ is the perturbation term of the journal part, T_ζ is the perturbation term of the thrust part. The perturbation of horizontal, vertical and axial displacement corresponds to the situation when ξ is equal to x, y, z respectively. Similarly, the perturbation of horizontal, vertical and axial velocity correspond to the situation when ξ is equal to $\dot{x}$, $\dot{y}$, $\dot{z}$ respectively. The stiffness and damping can be obtained after getting disturbance pressure by integration as the following equations show:

$$\boldsymbol{K}_J = \iint\limits_J \begin{Bmatrix} -\cos\theta \\ -\sin\theta \\ 0 \end{Bmatrix} \{p_x p_y p_z\} \mathrm{d}\Omega_J = \begin{bmatrix} Kxx & Kxy & Kxz \\ Kyx & Kyy & Kyz \\ 0 & 0 & 0 \end{bmatrix} \tag{2.24}$$

$$\boldsymbol{C}_J = \int\limits_J \begin{Bmatrix} -\cos\theta \\ -\sin\theta \\ 0 \end{Bmatrix} \{p_{\dot{x}} p_{\dot{y}} p_{\dot{z}}\} \mathrm{d}\Omega_J = \begin{bmatrix} Cxx & Cxy & Cxz \\ Cyx & Cyy & Cyz \\ 0 & 0 & 0 \end{bmatrix} \tag{2.25}$$

$$\boldsymbol{K}_T = \iint\limits_T \begin{Bmatrix} 0 \\ 0 \\ -1 \end{Bmatrix} \{p_x p_y p_z\} \mathrm{d}\Omega_T = \begin{bmatrix} 0 & 0 & 0 \\ 0 & 0 & 0 \\ Kzx & Kzy & Kzz \end{bmatrix} \tag{2.26}$$

$$\boldsymbol{C}_T = \iint\limits_T \begin{Bmatrix} 0 \\ 0 \\ -1 \end{Bmatrix} \{p_{\dot{x}} p_{\dot{y}} p_{\dot{z}}\} \mathrm{d}\Omega_T = \begin{bmatrix} 0 & 0 & 0 \\ 0 & 0 & 0 \\ Czx & Czy & Czz \end{bmatrix} \tag{2.27}$$

2.2.7 *Simulation Procedure*

Compared with the conventional bearing lubrication model, the proposed model needs to consider the journal and thrust parts, and their coupled effect. On the basis of lubrication calculation of conventional bearing, adding the temperature iterative calculation considering the heat flow continuity condition and the pressure iterative calculation considering the flow continuity condition is necessary. The calculation flow chart of the TEHD lubrication and dynamic characteristic parameters for journal-thrust coupled bearing is drawn in Fig. 2.4. At first, steady-state hydrodynamic pressure and oil film temperature distribution of single part are calculated. Then the coupled relationship between journal and thrust parts is established to update the oil film temperature of each part. To be specific, (1) the thermodynamic results are estimated through the calculation of the individual parts, (2) the results are substituted into the thermal continuity equation; (3) the coupled temperature distribution is calculated using the overrelaxation iteration. Finally, the pressure distributions considering coupled effect are calculated by the same method as temperature. The stiffness and damping in all directions are obtained based on the above lubrication results by using the perturbed Reynolds equation.

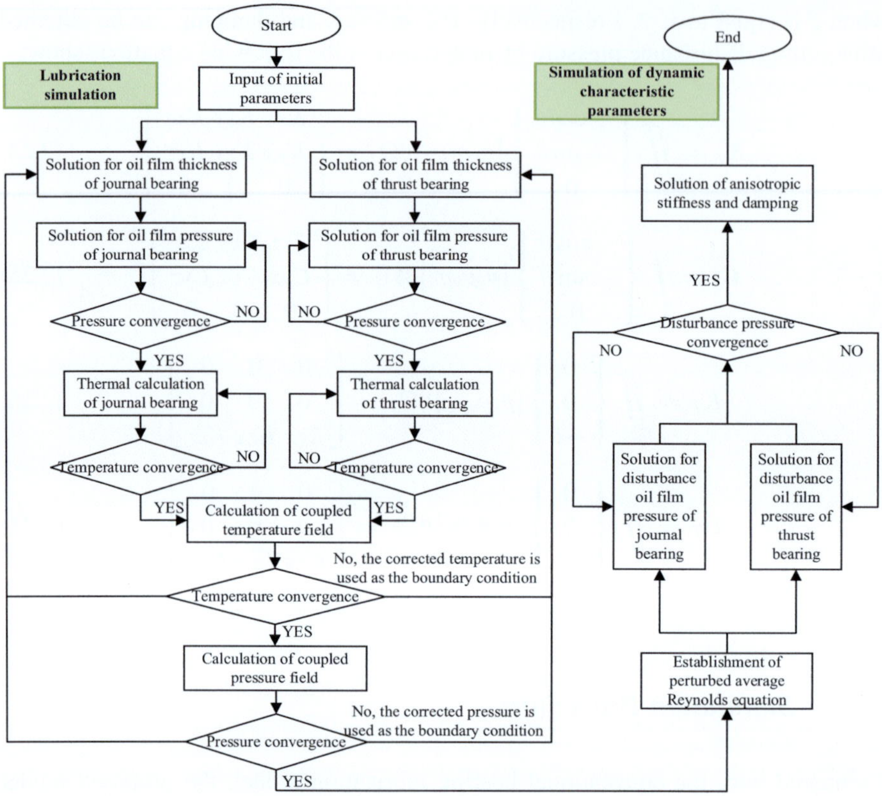

Fig. 2.4 Flow chart of the TEHD lubrication and dynamic model for journal-thrust coupled bearing

2.3 Analysis of the Steady-State Lubrication Characteristics of Journal-Thrust Coupled Bearings Under Constant Load

2.3.1 Model Verification

For verifying the correctness of the proposed model, this chapter set up a specific test rig and carried out the steady-state lubrication test. The proposed model was verified through comparison between the simulation results (including friction coefficient and film temperature) and the experimental results. Figure 2.5a shows the test rig and journal-thrust coupled bearing. Figure 2.5b shows a two-dimensional sketch of the entire test rig, including diagram which shows the arrangement of temperature measuring points on the bearing. Six temperature detecting devices were respectively installed on the surfaces of journal bearing (point 1–3) and thrust bearing (point 4–6) to obtain the bearing temperature. The measuring points 5 and 6 of the thrust surface were arranged in the leading edge to measure the inlet oil temperature of the thrust part, while point 4 was arranged in the trailing edge to measure the outlet oil temperature. The measuring points 2 of the journal surface was arranged in the main bearing area, while points 1 and 3 were arranged in both sides of bearing. The specific parameters related to the experiment can be found in Table 2.1. The friction coefficient could be calculated through friction torque obtained from the torque sensor. The axial force loading device on the left side and the radial force loading device on the bottom were used to provide stable loads to the bearing during the experiment. The temperature and friction coefficient were recorded constantly, and the stable value after 2 h of operation was taken as the steady-state result.

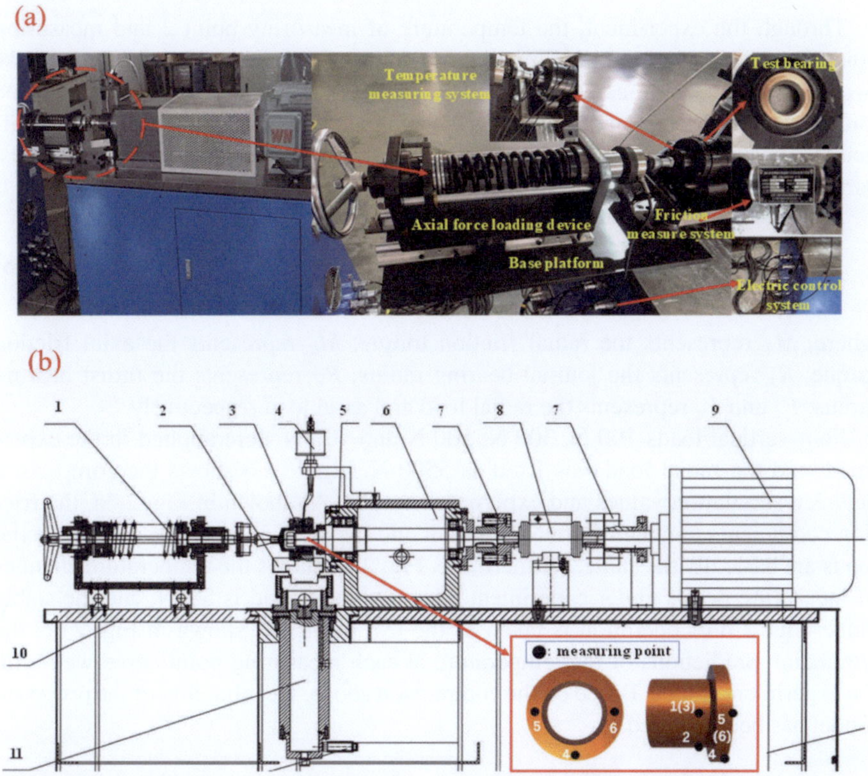

1- Axial force loading device; 2- Oil receiving box; 3- Test fixture; 4- Oil box; 5-test bearing; 6- Bearing block; 7- Coupling; 8- Torque sensor; 9-Motor; 10-Base platform; 11- Radial force loading device

Fig. 2.5 **a** Picture of the test rig; **b** Schematic of the test rig (including temperature measuring points distribution). 1- Axial force loading device; 2- Oil receiving box; 3- Test fixture; 4- Oil box; 5-test bearing; 6-Bearing block; 7- Coupling; 8- Torque sensor; 9-Motor; 10-Base platform; 11- Radial force loading device

Table 2.1 Experiment parameters

Parameters	Value
Internal diameter of journal part	40 mm
External diameter of thrust part	60 mm
Inclination angle	0.002°
Velocity/ω	500 r/min
Viscosity of the oil/η_0	0.175 Pa·s
Supply temperature/T_0	20 °C
Radial load	5500 N
Vertical load	100/300/500/700 N

Through the experiment, the temperature of measuring point 2 and measuring point 4 were found as the maximum temperature of journal and thrust parts. These two measuring points can clearly represent the temperature difference between journal and thrust parts and were selected for the comparison. The friction coefficient measured in the experiment was the sum of the journal part and the thrust part, which was given by

$$\mu = \frac{M_1}{F_y R_1} + \frac{2M_2}{F_Z(R_1 + R_T)} \tag{2.28}$$

where, M_1 represents the radial friction torque; M_2 represents the axial friction torque; R_I represents the journal bearing radius; R_T represents the thrust bearing radius; F_y and F_z represents the radial load and axial load respectively.

Four vertical loads 100 N, 300 N, 500 N and 700 N were applied in the experiment and the radial load was fixed as 5500 N. Figure 2.6 shows the comparison between simulation values and experiment results. As shown in Fig. 2.6a, the friction coefficients obtained from experiment and the simulation under different axial loads are basically the same. Furthermore, Fig. 2.6b shows the temperature changes of measuring points under experiment when the axial load is 500N, and the stable value after 2 h of operation is taken as the test result. As shown in Fig. 2.6c, the simulation predictions of the temperature at each measuring point agree well with the experiment results. Based on the comparison above, the reliability of the proposed model has been verified.

2.3.2 *Effect of Journal-Thrust Coupled Lubrication Effect on Lubrication Characteristics*

In order to analyze the necessity of thermal and pressure coupled effects in the lubrication process of journal-thrust coupled bearing, this section carried out the steady-state lubrication analysis with the fixed eccentricity ratio and the fixed thrust geometry

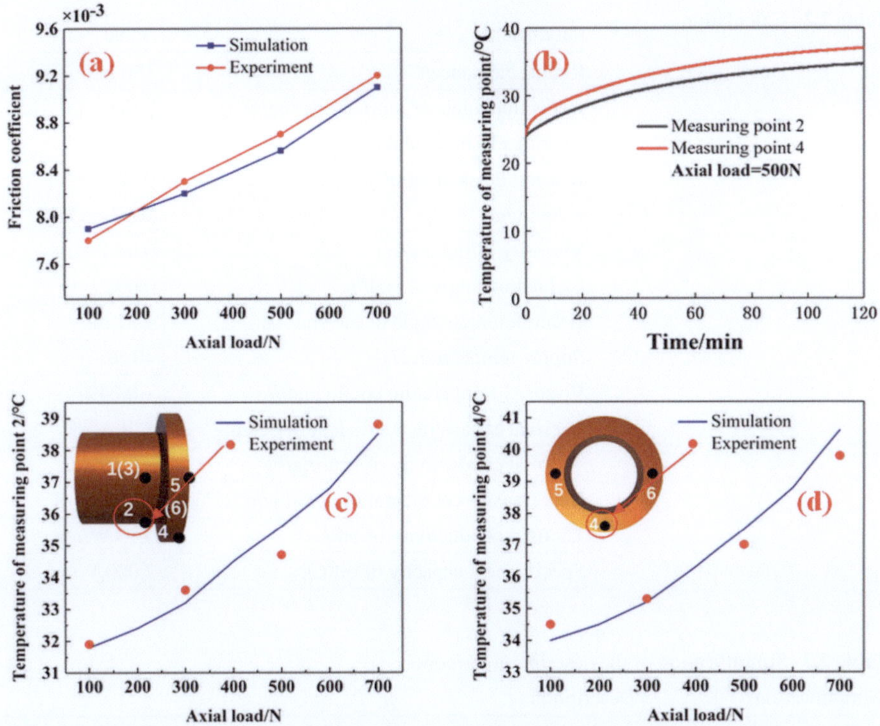

Fig. 2.6 **a** Verification of the friction coefficient; **b** Temperature of the measuring points under test; **c** Verification of the temperature in points 2; **d** Verification of the temperature in points 4

film thickness. Bearing studied in this section had dimple texture on the surface of thrust part. In addition, the TEHD lubrication model without considering coupled effect (Without coupled model), the EHD lubrication model considering pressure coupled effect (EHD coupled model), and the TEHD lubrication model considering thermal and pressure coupled effects (TEHD coupled model) were respectively used for comparative analysis. Table 2.2 shows the calculation parameters of the journal-thrust coupled bearing of diesel engine used in this chapter. In order to make the description of models more concise, some simplifications are introduced as Table 2.3 shows.

Figure 2.7 reveals the pressure distribution of journal-thrust coupled bearing calculated by different models when the eccentricity ratio is 0.8 (journal part) and the geometry clearance is 10 μm (thrust part). In general, compared with the results of "without coupled model", the peak pressure of "TEHD coupled model" increases from 11.75 to 14.4 MPa (an increase of about 23%) in journal part, and increases from 15.25 MPa to 19.16 MPa (an increase of about 26%) in thrust part, along with significant changes in distribution. Comparing Fig. 2.7a, c, the pressure of the journal part shifts to the thrust side after considering thermal and pressure coupled effects, accompanied by four peak pressures appear on the axial section. In addition, the axial

Table 2.2 Calculation parameters

Parameters	Value
Radius clearance/C	150 μm
Deflection angle of journal bearing/φ	0 rad
Bearing elastic modulus/E_J	105 GPa
Bearing Poisson ratio/v_p	0.327
velocity/ω	36.65 rad/s
Viscosity of the oil/η_0	0.07 Pa·s
Inclination angle of pad/θ_p	0.002°
circumferential angle of the Pad /θ_0	1.31 rad
Supply temperature/T_0	40 °C
Viscosity temperature coefficient/β	−0.0491
Thermal expansion coefficient of bearing/k_p	1.2E−5/K
Thermal conductivity of bearing/k_b	47 W/mK
Heat transfer coefficient of bearing/h_b	80 W/m^2 K
Thermal conductivity of oil/k_r	0.13 W/mK
Specific heat capacity of oil/Cp_o	2,000 W/m^2 K

Table 2.3 Simplification of the description of models

Simplification	Description
Without coupled model	TEHD lubrication model without considering coupled effect
EHD coupled model	EHD lubrication model considering pressure coupled effect
TEHD coupled model	TEHD lubrication model considering thermal and pressure coupled effects

pressure is distributed asymmetrically, and the maximum pressure starts to shift to the thrust side due to the coupled effect. By comparing Fig. 2.7d, f, the pressure area of the thrust part expands towards the common boundary, and the maximum pressure appears in the circumferential middle position (position of peak pressure in journal part). This is because that, after considering the continuity conditions, ending leakage effect in common boundary interface is weakened, while the pressure transmission between the axial direction in journal part and the radial direction in thrust part is enhanced.

Furthermore, compared with the results of "EHD coupled model", the peak pressure of "TEHD coupled model" increases from 13 MPa to 14.4 MPa (an increase of about 11%) in journal part, and increases from 17.4 to 19.16 MPa (an increase of about 10%) in thrust part, along with significant difference of the pressure distribution in high-pressure area. By comparing Fig. 2.7b, c, the pressure distribution of journal part in the circumferential middle position changes obviously. After considering the thermal coupled effect, the value of maximum pressure and area of the high-pressure zone in journal part both increase. This is because the corresponding

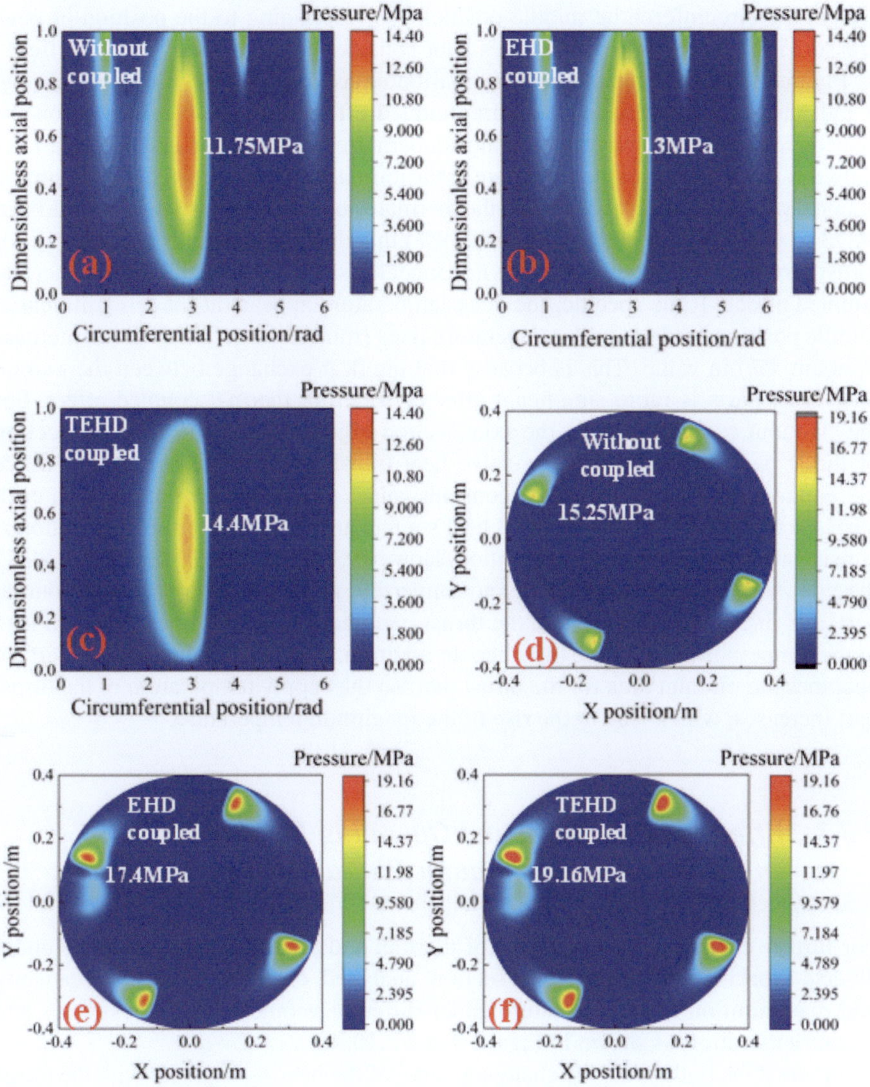

Fig. 2.7 Pressure distribution of journal-thrust coupled bearing: **a** Journal part (Without coupled); **b** Journal part (EHD coupled); **c** Journal part (TEHD coupled); **d** Thrust part (Without coupled); **e** Thrust part (EHD coupled); **f** Thrust part (TEHD coupled)

temperature of the high-pressure zone would rise after considering the thermal effect, which can lead to greater thermal expansion, smaller film thickness, and eventually higher hydrodynamic pressure. By comparing Fig. 2.7e, f, the change of pressure distribution of thrust part is similar to that of the journal part, and the high-pressure area at the circumferential middle position (corresponding to the position of peak pressure in the journal part) increases after considering the thermal coupled effect.

Figure 2.8 reveals the temperature distribution of journal-thrust coupled bearing. It indicates that thermal and pressure coupled effects has great influences on the distribution and maximum values of the temperature of journal part and thrust part. As shown in Fig. 2.8a, b, peak temperature of the journal part moves towards the common boundary interface after considering the thermal coupled effect, which increases from 44.73 to 48.78 °C (an increase of about 9%) in value. As shown in Fig. 2.8c, d, for thrust part, temperature of each pad is no longer the same after considering the thermal coupled effect. To be specific, the peak temperature appears at the circumferential middle position, and the peak temperature rises from 56.12 to 58.37 °C (an increase of about 4%) in value. This is because that the heat exchange between the journal and thrust parts is more significant after considering thermal coupled effect. For the "without coupled model", the axial section of journal part and the inner section of thrust part both undergo convective heat transfer with the environment. Since the environment temperature is a constant value, the oil film temperature in each part is greatly influenced by internal heat source, and the temperature distribution is consistent with the pressure distribution. However, for the "TEHD coupled model", the common boundary interface is no longer the oil drainage area for the journal part, and the heat exchange with the thrust part needs to be considered, which lead to the temperature rise of the interface. In addition, the common boundary interface becomes the oil inlet area for the thrust part, so the supply temperature of the thrust part increases, which lead to the rise of the maximum temperature.

2.3.3 *Effects of Structural and Operating Parameters on Steady-State Lubrication Characteristics*

For further exploring the influences of thermal and pressure coupled effects under diverse working conditions, this section analyzed changes of bearing capacity and maximum oil film temperature under different geometry film thicknesses and eccentricity ratios, as shown in Figs. 2.9 and 2.10.

Figure 2.9a, b illustrate the changing curve of the bearing capacity with the thrust part geometry film thickness (h_p) under three eccentricity ratios (0.7, 0.8 and 0.9). The "without coupled model" and the "EHD coupled model" were also added in Fig. 2.9 for comparison. Figure 2.9a shows that, with the decrease of h_p, the bearing capacity of the journal part decreases after considering the coupled effect. However, the bearing capacity does not change with h_p when coupled effect is not considered, while it is still lower than the results under other conditions. This is because the film pressure of the thrust part increases significantly with the reduction of h_p, which cause partly rise of the pressure at common boundary interface. However, pressure at other positions in the journal part decreases slightly, thus ultimately result in bearing capacity of the entire journal part decreasing. In addition, by comparing the results

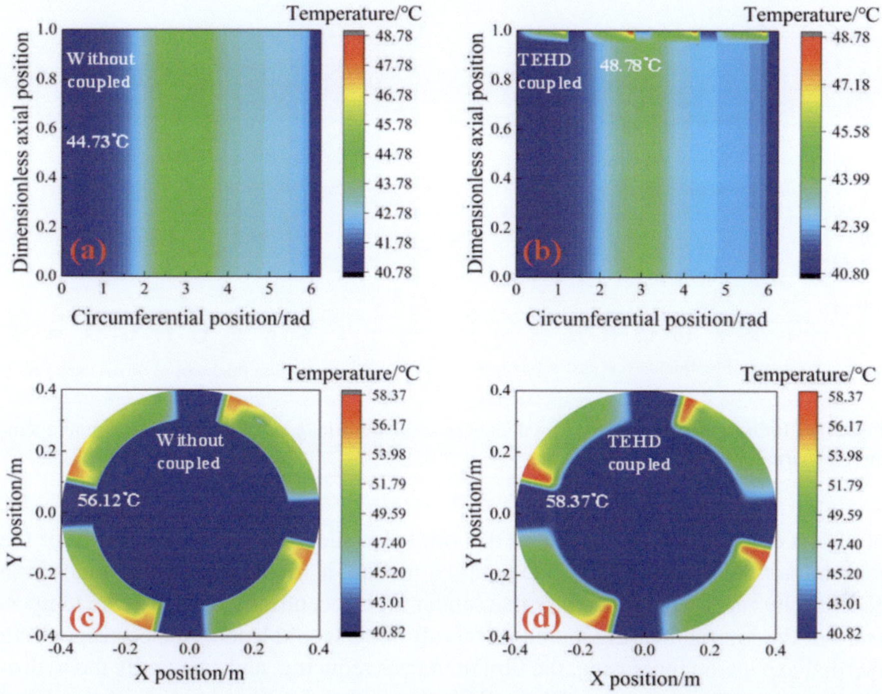

Fig. 2.8 Temperature distribution of journal-thrust coupled bearing: **a** Journal part(Without coupled); **b** Journal part(TEHD coupled); **c** Thrust part(Without coupled); **d** Thrust part(TEHD coupled)

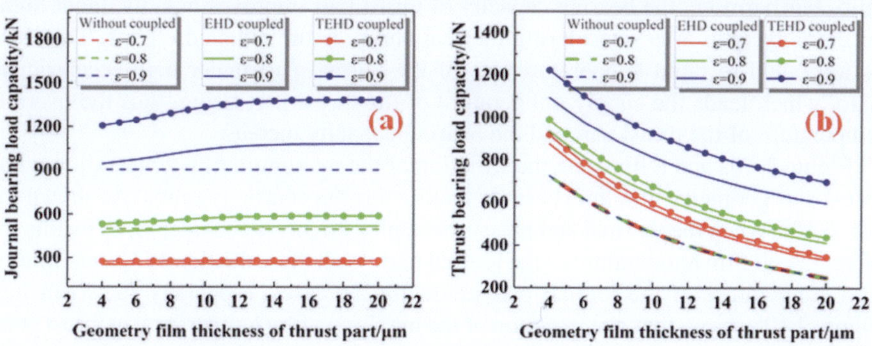

Fig. 2.9 Influences of geometry film thickness and eccentricity ratio on the bearing capacity: **a** Journal part; **b** Thrust part

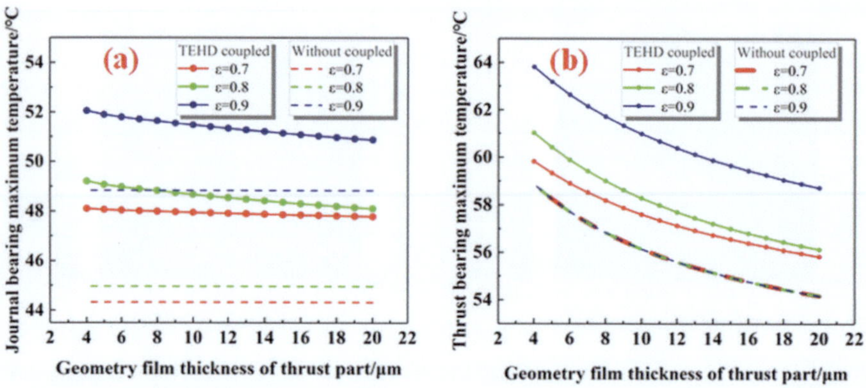

Fig. 2.10 Influences of geometry film thickness and eccentricity ratio on the maximum temperature: **a** Journal part; **b** Thrust part

of "EHD coupled model" and "TEHD coupled model", the bearing capacity of the journal part calculated by "TEHD coupled model" is higher, especially under high eccentricity ratio. This is because that, under high eccentricity ratio, oil film temperature of the journal bearing increases rapidly. This would lead to the corresponding thermal expansion increasing, the film thickness reducing, and eventually the hydrodynamic pressure increasing. Figure 2.9b shows that, bearing capacity of the thrust part rises significantly with the decrease of h_p, especially at small h_{p}. In addition, because of the consideration of coupled effect, the bearing capacity of the thrust part gradually rises along with the eccentricity ratio, especially at high eccentricity ratio. Furthermore, the bearing capacity of thrust part improves notably under high eccentricity ratio after considering thermal and pressure coupled effects. This can be attributed to rapid temperature rise of the journal part under high eccentricity ratio, which leads the supply temperature of the thrust part rising, thus the overall temperature of the thrust part and the bearing capacity increase.

Figure 2.10a and b illustrate the variation of the maximum film temperature with thrust part geometry film thickness (h_{p}) under three eccentricity ratios. As shown in Fig. 2.10a, when the thermal and pressure coupled effects is considered, h_{p} begins to affect the oil film temperature of the journal part. With the reduction of h_{p}, maximum film temperature of the journal part gradually increases, especially for small h_p. Figure 2.10b shows that, the variation of the maximum film temperature in thrust part with h_{p} is similar to that in journal part. Moreover, the maximum film temperature in thrust part rises steadily as the eccentricity ratio increases. The variation of maximum film temperature is consistent with that of bearing capacity illustrated in Fig. 2.9. This is because that, the oil film temperature of journal part rises rapidly at high eccentricity ratio, leading to a significant increase in the supply temperature of the thrust part.

2.4 Analysis of the Steady-State Dynamic Characteristic Coefficients of Journal-Thrust Coupled Bearings Under Constant Load

2.4.1 Literature Verification

Based on the degenerated version of the proposed model, this chapter also verified the stiffness and damping of journal bearing (Liu et al. 2022) and thrust bearing (Wang et al. 2018) respectively. Figure 2.11a, b show the comparison between simulation values and verification values of journal bearing and thrust bearing respectively. Input parameters used in the calculation were given by the corresponding literature. Results calculated from the degradation model match well with the literature results (the average error less than 10%), which confirms the correctness of the proposed model.

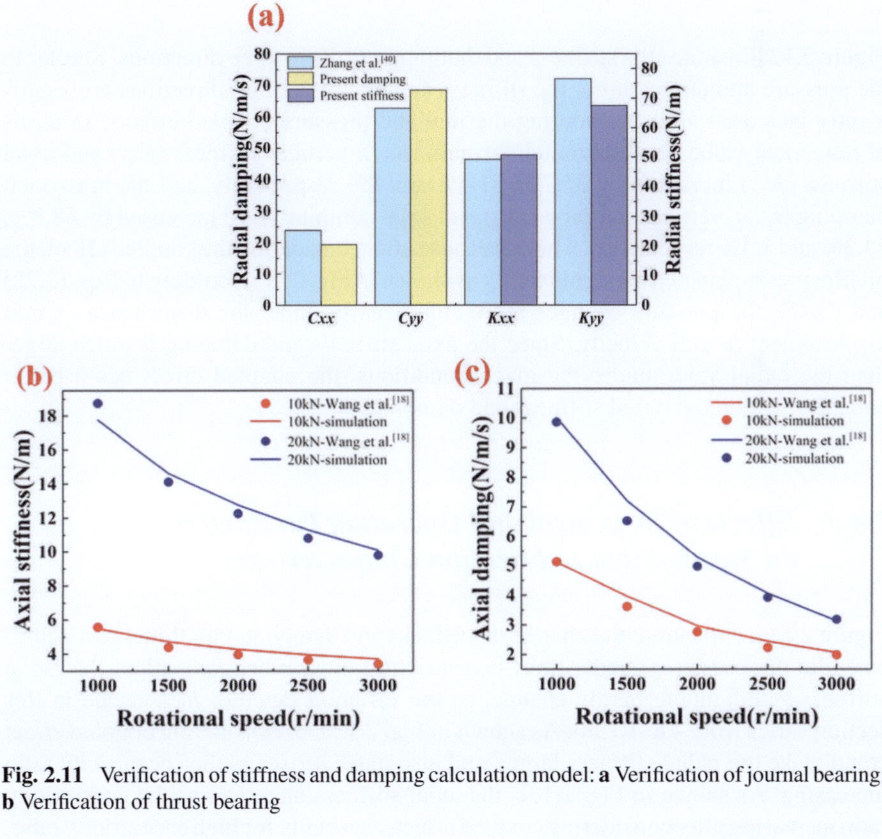

Fig. 2.11 Verification of stiffness and damping calculation model: **a** Verification of journal bearing; **b** Verification of thrust bearing

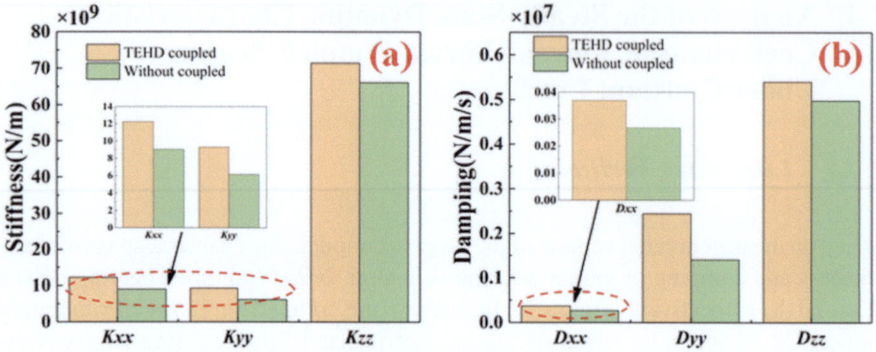

Fig. 2.12 Stiffness and damping of journal-thrust coupled bearing: **a** Stiffness; **b** Damping

2.4.2 Effect of Journal-Thrust Coupled Lubrication Effect on Lubrication Characteristics

Figure 2.12 illustrates the stiffness and damping results in three directions. Similar to the pressure and temperature, the stiffness and damping of all directions are significantly increased after considering thermal and pressure coupled effects. In terms of numerical value, the horizontal stiffness (K_{xx}), vertical stiffness (K_{yy}) and axial stiffness (K_{zz}) increased by 35.7%, 51.4% and 8% respectively, and the horizontal damping (D_{xx}), vertical damping (D_{yy}) and axial damping (D_{zz}) increased by 38.3%, 74.3% and 8.4% respectively. This is because after considering the coupled effect, the oil film pressure increases significantly as shown in Fig. 2.7. According to Eqs. (2.22) and (2.23), the pressure changes more significantly under the disturbance of unit displacement or unit velocity. Since the axial stiffness and damping is much larger than the radial value under the given conditions, the coupled effect has a better enhancement on the radial stiffness and damping.

2.4.3 Effects of Structural and Operating Parameters on Steady-State Lubrication Characteristics

Figure 2.13a, c illustrate the change of stiffness and damping with thrust part geometry film thickness (h_p) under three eccentricity ratios. When h_p is high, the radial stiffness and damping hardly change, so the variation range of h_p selected in this section varies from 4 to 10 μm. As shown in Fig. 2.13a, b, considering coupled effect could make the radial stiffness larger, and also make it rises as the eccentricity ratio increasing. As shown in Fig. 2.13c, the axial stiffness also rises as the eccentricity ratio increasing after considering coupled effect, especially for high eccentricity ratio. In addition, because of the consideration of thermal and pressure coupled effects,

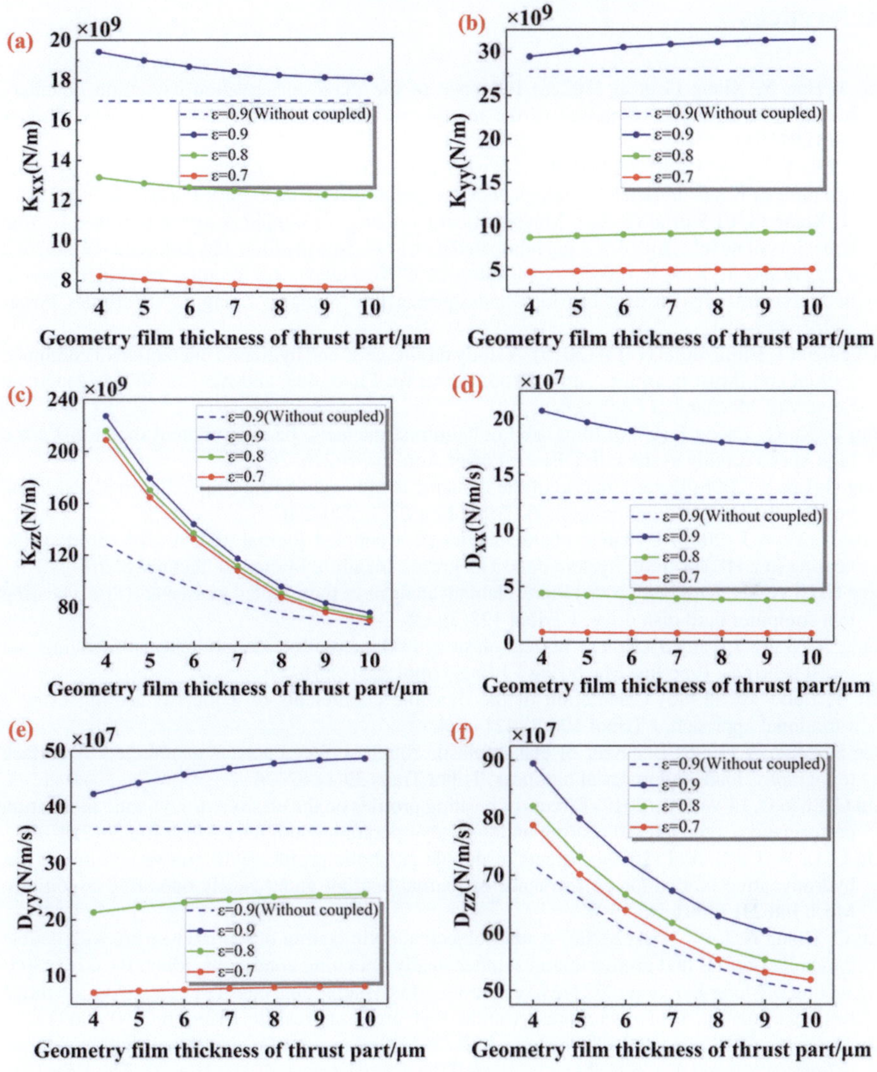

Fig. 2.13 Influences of geometry film thickness and eccentricity ratio on the stiffness and damping: **a** K_{xx}; **b** K_{yy}; **c** K_{yy}; **d** D_{xx}; **e** D_{yy}; **f** D_{zz}

the radial stiffness no longer remains unchanged with the reduction of h_p. Axial and horizontal stiffness both increase with the decrease of h_p, while vertical stiffness decreases. As shown in Fig. 2.13e, g, the variation of damping is consistent with the stiffness. The changing of stiffness and damping is similar to that of bearing capacity, which can be explained by Eqs. (2.18) and (2.21).

References

Cai J, Han Y, Xiang G et al (2022a) Influence of the mass conservation cavitation boundary on the tribo-dynamic responses of the micro-groove water-lubricated bearing. Surf Topogr 10(4):045011

Cai J, Han Y, Xiang G et al (2022b) Effects of wear and shaft-shape error defects on the tribo-dynamic response of water-lubricated bearings under propeller disturbance. Phys Fluids 34(7):077118

Cai J, Xiang G, Li S et al (2022c) Mathematical modeling for nonlinear dynamic mixed friction behaviors of novel coupled bearing lubricated with low-viscosity fluid. Phys Fluids 34(9):093612

Guo Z, Hirano T, Kirk R (2005) Application of CFD analysis for rotating machinery part I: hydrodynamic, hydrostatic bearings and squeeze film damper. J Eng Gas Turbines Power 127(2):445–451

Hagemann T, Blumenthal H et al (2015). A study on energetic and hydraulic interaction of combined journal and thrust bearings. Turbo Expo: Power for Land, Sea, and Air. vol 56765. American Society of Mechanical Engineers

Han B, Xu Q, Zheng S (2016) Integrated radial/thrust magnetic bearing without thrust disk for a high-speed driving system. IET Electr Power Appl 10(4):276–283

Jang G, Lee S (2006) Determination of the dynamic coefficients of the coupled journal and thrust bearings by the perturbation method. Tribol Lett 22(3):239–246

Jang G, Yoon J (2002) Dynamic characteristics of a coupled journal and thrust hydrodynamic bearing in a HDD spindle system due to its groove location. Microsyst Technol 8:261–270

Jang GH, Lee SH, Kim HW (2006) Finite element analysis of the coupled journal and thrust bearing in a computer hard disk drive. J Tribol 128(2):335–340

Jiang X, Wang J, Fang J (2011) Thermal elastohydrodynamic lubrication analysis of tilting pad thrust bearings. Proc Inst Mech Eng J J Eng Tribol 225(2):51–57

Klit P, Lund JW (1986) Calculation of the dynamic coefficients of a journal bearing using a variational approach. J Tribol 108(3):421–424

Lee SC, Ren N (1996) Behavior of elastic-plastic rough surface contacts as affected by surface topography, load, and material hardness. Tribol Trans 39(1):67–74

Liu C, Zhao B, Li W et al (2019) Effects of bushing profiles on the elastohydrodynamic lubrication performance of the journal bearing under steady operating conditions. Mech Ind 20(2):1–13

Liu C, Li W, Lu X et al (2020a) Effects of double parabolic profiles with groove textures on the hydrodynamic lubrication performance of journal bearing under steady operating conditions. Mech Ind 21(3):301

Liu C, Zhong N, Lu X et al (2020b) A multiobjective optimization of journal bearing with double parabolic profiles and groove textures under steady operating conditions. Mech Ind 21(3):305

Liu R, Jing L, Meng X, Lyu B (2021) Mixed elastohydrodynamic analysis of a coupled journal-thrust bearing system in a rotary compressor under high ambient pressure. Tribol Int 159:106943

Liu Q, Ouyang W, Cheng Q, Li J, Cheng Q, Li R (2022) Influences of bidirectional shaft inclination on lubrication and dynamic characteristics of the water-lubricated stern bearing. Mech Syst Sig Process 169:108623

Majumdar B, Pai R, Hargreaves D (2004) Analysis of water-lubricated journal bearings with multiple axial grooves. Proc Inst Mech Eng J J Eng Tribol 218(2):135–146

Pai RS, Pai R (2008) Stability of four-axial and six-axial grooved water-lubricated journal bearings under dynamic load. Proc Inst Mech Eng J J Eng Tribol 222(5):683–691

Patir N, Cheng H (1978) An average flow model for determining effects of three- dimensional roughness on partial hydrodynamic lubrication. Trans ASME J Lubr Technol 100(1):12–17

Peixoto TF, Cavalca KL (2019) Investigation on the angular displacements influence and nonlinear effects on thrust bearing dynamics. Tribol Int 131:554–566

Shi F, Wang Q (1998a) A method of influence functions for thermal analyses of tribological elements. Tribol Trans 41(3):350–358

Shi F, Wang Q (1998b) A mixed-TEHD model for journal-bearing conformal contacts—part I: model formulation and approximation of heat transfer considering asperity contact. J Tribol 120(2):198–205

Stefani F (2006) THD analysis of a coupled journal and tilting pad thrust bearing system. Proceedings of the 5th international tribology conference, pp 20–22

Sun F, Zhang X, Wei Y, Wang X, Wang D (2020) Stability analysis of rubber-supported thrust bearing in a rotor-bearing system used in marine thrusters under disturbing moments. Tribol Int 151(4):106356

Tian L, Wang WJ, Peng ZJ (2012) Effects of bearing outer clearance on the dynamic behaviours of the full floating ring bearing supported turbocharger rotor. Mech Syst Sig Process 31:155–175

Tian L, Wang WJ, Peng ZJ (2013) Nonlinear effects of unbalance in the rotor-floating ring bearing system of turbochargers. Mech Syst Sig Process 34(1–2):298–320

Tucker PG, Keogh PS (1995) A generalized computational fluid dynamics approach for journal bearing performance pre-diction. Proc Inst Mech Eng J J Eng Tribol 209(2):99–108

Wang Y, Wang Q et al (2002) A mixed-TEHD analysis and experiment of journal bearings under severe operating conditions. Tribol Int 35(6):395–407

Wang Y, Wang Q, Lin C (2003) Mixed lubrication of coupled journal-thrust-bearing systems including mass conserving cavitation. J Tribol 125(4):747–755

Wang W, Liu B, Zhang Y, Shao X, Allaire P (2018) Theoretical and experimental study on the static and dynamic characteristics of tilting-pad thrust bearing. Tribol Int 123(1):26–36

Wang X, Adachi K, Otsuka K, Kato K (2006) Optimization of the surface texture for silicon carbide sliding in water. Appl Surf Sci 253(3):1282–1286

Xiang G, Han Y, Chen R, Wang J, Ni X (2019) A numerical method to investigate the mixed lubrication performances of journal-thrust coupled bearings. Idn Lubr Tribol 71(9):1099–1107

Xiang G, Wang J, Zhou C, Shi Y, Wang Y, Cai J et al (2021) A tribo-dynamic model of coupled journal-thrust water-lubricated bearings under propeller disturbance. Tribol Int 160:107008

Xiang G, Yang T, Guo J et al (2022) Optimization transient wear and contact performances of water-lubricated bearings under fluid-solid-thermal coupling condition using profile modification. Wear 502–503:204379

Xie Z, Zhang Y, Zhou J, Zhu W (2021) Theoretical and experimental research on the micro interface lubrication regime of water lubricated bearing. Mech Syst Sig Process 151:107422

Xie Z, Jiao J, Wrona S (2023a) The fluid-structure interaction lubrication performances of a novel bearing: experimental and numerical study. Tribol Int 179:108151

Xie Z, Jiao J, Yang K (2023b) Theoretical and experimental study on the fluid-structure-acoustic coupling dynamics of a new water lubricated bearing. Tribol Int 177:107982

Xu H, He T, Zhong N et al (2022) Transient thermomechanical analysis of micro cylindrical asperity sliding contact of SnSbCu alloy. Tribol Int 167:107362

Zhang L, Zhang F, Qin Z et al (2022) Piezoelectric energy harvester for rolling bearings with capability of self-powered condition monitoring. Energy 238:121770

Chapter 3
Transient Lubrication and Dynamic Characteristics Analysis of Crankshaft Journal-Thrust Coupled Bearings Under Real Engine Load Conditions

Abstract The Journal-Thrust coupling bearing (JTC bearing) operates at the end of the diesel engine crankshaft and has the functions of bearing radial loads and preventing axial movement of the crankshaft. In this paper, the Transient Thermoelastic Fluid Dynamics (TEHD) lubrication model is adopted to analyze the JTC bearings of Marine diesel engines. Based on the flow rate, pressure and thermal continuity conditions between the journal and the thrust component, the thermocompression coupling effect was considered. The transient time-varying load was considered in the lubrication analysis, and the coupling effect under the actual working conditions was discussed. A transient TEHD lubrication experimental platform for JTC bearings was established, and the established model was verified. It is found that the thermocompressive coupling effect between the journal and the thrust component has a significant influence on the transient lubrication behavior of JTC bearings and cannot be ignored. The coupling influence of different boundary conditions on transient lubrication characteristics such as axial clearance, radial clearance, radial load waveform, inclination Angle and inlet oil temperature was studied. This chapter provides theoretical support for the lubrication design of JTC bearings used in diesel engine crankshafts.

Keywords Transient analysis · Thermal elastohydrodynamic lubrication · Thermal-pressure coupled effect · Dynamic load

3.1 Introduction

Journal-thrust coupled bearing (abbreviated as JTC bearing) is a specialized type of bearing positioned at the end of the crankshaft, encompassing both the journal and thrust components. The journal part of this distinctive bearing primarily handles radial forces to provide support for the crankshaft, while the thrust part is designed to counteract axial forces, preventing axial displacement of the crankshaft. Unlike those two conventional bearings, the JTC bearing combines their functions. Due to its unique integrated structure, the oil supply for the thrust part depends on the oil

B. Zhao et al., *Lubrication and Dynamic Characteristics Analysis Technology*,
https://doi.org/10.1007/978-981-95-2270-5_3

leakage from the journal part, establishing inherent connections in the lubrication of these two parts. Researchers have explored this topic and achieved notable results. In the realm of industrial machinery lubrication research for JTC bearings, Wang et al. (Wang et al. 2002, 2003) identified a coupled lubrication relationship between the journal and thrust parts. They derived a continuity equation based on the continuous flow characteristics of the lubricant. While as an initial mechanism study, the model only considered steady-state operating conditions and hydrodynamic lubrication. Subsequently, Stefani (2006) built a test rig for coupled bearings and found the coupled phenomenon during the heat exchange of lubricant. Thus they established a steady-state thermal hydrodynamic model considering thermal continuity, while the thermal deformation was ignored in this work. Building upon prior research, Xiang et al. (2019) further improved the model by considering elastic deformation, and studied the lubrication performances of JTC bearing under the water-lubricated condition. The authors (Shi et al. 2023) discovered the thermal and pressure (thermal-pressure) coupled effect between the journal and thrust parts in previous studies, and upgraded the model to a steady-state TEHD model. Relevant analysis showed that the eccentricity and axial clearance both had influences on the journal and thrust parts when considering the thermal-pressure coupled effect. Xiang et al. (2021a, b) considered propeller-induced excitation and upgraded a steady-state EHD lubrication model to a transient EHD lubrication model in subsequent research. They explored the dynamic characteristics under propeller disturbance, focusing on the stern bearing features within the propulsion shafting, while the thermal effect was ignored since the temperature rise was not so notable on that occasion.

The existing research on the lubrication behaviors of JTC bearings appears to be less comprehensive compared to traditional journal or thrust bearings. Many of the current models in this domain predominantly address steady-state conditions, often overlooking transient operating states and thermal effects in lubrication, particularly concerning coupled bearings used in diesel engines. However, under real-world operational circumstances, diesel engine crankshafts experience time-varying loads, resulting in intricate and challenging lubrication conditions for coupled bearings. Additionally, as lubricant temperatures rise, the thermal effect becomes increasingly significant and cannot be disregarded. Consequently, it is imperative to establish a transient Thermal Elastohydrodynamic (TEHD) model for JTC bearings under actual diesel engine working conditions. Moreover, further exploration of the thermal-pressure coupled effect between the journal and thrust parts is essential, especially under transient loads.

In this chapter, a transient TEHD model for the JTC bearing in the marine diesel engine was established to explore its lubrication characteristics under time-varying load condition. Based on the established model, influences of the transient coupled effects under different input parameters were analyzed, including axial clearances, waveforms of radial load, incline angles, inlet oil temperatures and radial clearances. The findings aim to offer valuable insights for guiding the design of JTC bearings in marine diesel engines. Furthermore, in order to better simulate the actual machine working conditions, the actual machine operating conditions were selected for research. The influences of different structure, operating condition and lubricant parameters on the tribology performance and dynamic characteristic were investigated. The time-varying coupled lubrication mechanism was revealed, and the evolution law of the dynamic characteristics of the JTC bearing was explored. On this basis, the natural characteristics of the crankshaft system were further analyzed to support the subsequent investigation into its vibration properties of the crankshaft system. This study can serve as a foundational step for examining the dynamic characteristics of JTC bearings under real operating conditions, and offer insights for the development of JTC bearings.

3.2 Transient-State Lubrication and Dynamic Characteristics Analysis Model Considering Journal-Thrust Coupled Lubrication Effect

3.2.1 Lubrication Model

Figure 3.1 shows schematic diagram of a JTC bearing. Figure 3.1a and b show the front view and the side view of journal part, taking into account the inclination angle of the journal. Similarly, Fig. 3.1c and d show the front view and the side view of the thrust part, where the tilt of the journal creates a wedge-shaped gap between the thrust collar and the bearing. γ is the inclination angle of the journal, R_2 and R_1 represent the outer and inner radius of thrust part, h_0 represents the geometry clearance of the thrust part.

The film thickness of the journal part h_J can be represented as:

$$h_J = c + \mathrm{e}\cos(\theta - \varphi) + \tan\gamma\cos(\theta - \varphi)\left(Y - \frac{BL}{2}\right) + h_{JE} + h_{JT} \tag{3.1}$$

Similarly, the film thickness of the thrust part h_T can be expressed as:

$$h_T = h_0 + r\sin\gamma\cos\theta + h_{TE} + h_{TT} + h_s(t) \tag{3.2}$$

where $h_s(t)$ represents the axial shifting displacement affected by the instantaneous vibration of the crankshaft. This work used the axial shifting displacement tested under real operating condition.

The Reynolds equation (Liang et al. 2022) of the journal part and thrust part can be represented as follow:

$$\begin{aligned}\frac{\partial}{R_1^2\partial\theta}\left(\phi_\theta\frac{h_J^3}{\eta}\frac{\partial P_J}{\partial\theta}\right) + \frac{\partial}{\partial z}\left(\phi_z\frac{h_J^3}{\eta}\frac{\partial P_J}{\partial z}\right) \\ = \frac{6U_J}{R_1}\left(\phi_c\frac{\partial h_J}{\partial\theta} + \sigma\frac{\partial\varphi_s}{\partial\theta}\right) + 12\frac{\partial h_J}{\partial t}\end{aligned} \tag{3.3}$$

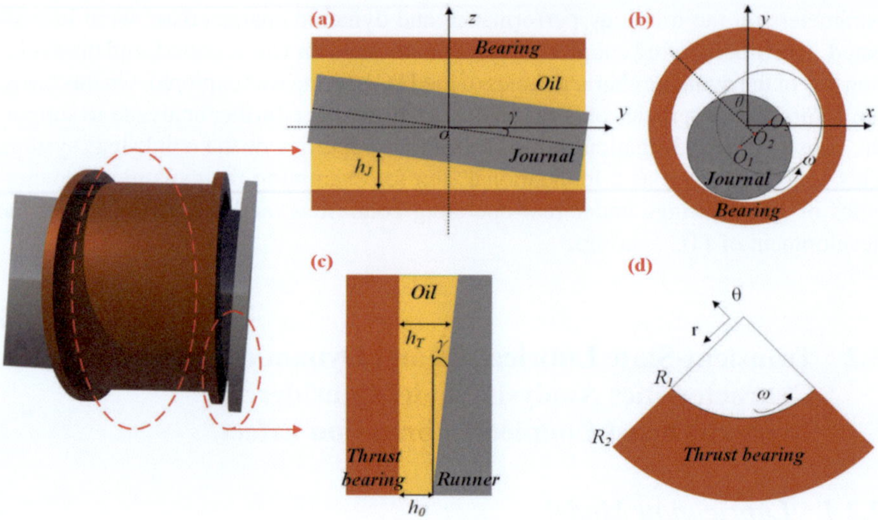

Fig. 3.1 Schematic diagram of the JTC bearing: **a** front view of the journal part; **b** side view of the journal part; **c** front view of the thrust part; **d** side view of the thrust part

$$\frac{\partial}{r\partial\theta}\left(\phi_\theta\frac{h_T^3}{\eta}\frac{\partial p_T}{\partial\theta}\right)+\frac{\partial}{\partial r}\left(\phi_r\frac{rh_T^3}{\eta}\frac{\partial p_T}{\partial r}\right)$$
$$=6\omega r\left(\phi_c\frac{\partial h_T}{\partial\theta}+\sigma\frac{\partial\phi_s}{\partial\theta}\right)+12\frac{\partial h_T}{\partial t} \tag{3.4}$$

The viscosity of the oil was calculated through the Reynolds viscosity formula (Li et al. 2019) as follows:

$$\eta=\eta_0\exp(\beta(T-T_0)) \tag{3.5}$$

Considering that the thermal effect of the JTC bearing is significant, the film temperatures of the two parts were calculated through the transient energy equation:

$$\rho C_{pl}\left(\frac{v_J}{r}\frac{\partial T_J}{\partial\theta}+w_J\frac{\partial T_J}{\partial z}+\frac{\partial T_J}{\partial t}\right)=\alpha T_J\left(w_J\frac{\partial P_J}{\partial z}+v_J\frac{\partial P_J}{r\partial\theta}\right)$$
$$+k_l\frac{\partial^2 T_J}{\partial r^2}+\eta\left[\left(\frac{\partial u_J}{\partial z}\right)^2+\left(\frac{\partial v_J}{\partial z}\right)^2\right] \tag{3.6}$$

$$\rho C_{pl}\left(U_r\frac{\partial T_T}{\partial r}+\frac{U_\theta}{r}\frac{\partial T_T}{\partial\theta}+\frac{\partial T_T}{\partial t}\right)=\alpha T_T\left(U_r\frac{\partial P_T}{\partial r}+U_\theta\frac{\partial P_T}{r\partial\theta}\right)$$
$$+k_l\frac{\partial^2 T_T}{\partial z^2}+\eta\left[\left(\frac{\partial U_r}{\partial z}\right)^2+\left(\frac{\partial U_\theta}{\partial z}\right)^2\right] \tag{3.7}$$

Assuming the bearing material is isotropic and the thermal conductivity, specific heat, density, and other material properties are constant, the heat conduction equations within the bearing are as follow:

$$\begin{cases} k_b\left(\dfrac{\partial^2 T_J}{\partial x^2} + \dfrac{\partial^2 T_J}{\partial y^2} + \dfrac{\partial^2 T_J}{\partial z^2}\right) \\ = c_{pb}\rho_b \dfrac{\partial T_J}{\partial t} \quad \textit{journal part} \\ k_b\left(\dfrac{\partial^2 T_T}{\partial r^2} + \dfrac{1}{r}\dfrac{\partial T_T}{\partial r} + \dfrac{1}{r^2}\dfrac{\partial^2 T_T}{\partial \theta^2} + \dfrac{\partial^2 T_T}{\partial z^2}\right) \\ = c_{pb}\rho_b \dfrac{\partial T_T}{\partial t} \quad \textit{thrust part} \end{cases} \tag{3.8}$$

The boundary conditions used in the thermal analysis of the journal part and the thrust part are as follows, where n is the unit vector in the convection heat transfer direction:

$$\begin{cases} k_b \frac{\partial T_J}{\partial n} = -h_b(T_J - T_\infty) & \textit{bearing-environment} \\ k_l \frac{\partial T_J}{\partial r} = k_b \frac{\partial T_J}{\partial r} & \textit{bearing-lubricant} \\ k_J \frac{\partial T_J}{\partial r} = \frac{1}{2\pi} k_l \int_0^{2\pi} \frac{\partial T_J}{\partial r} \mathrm{d}\theta & \textit{journal-lubricant} \end{cases} \tag{3.9}$$

$$\begin{cases} k_b \frac{\partial T_T}{\partial n} = -h_b(T_T - T_\infty) & \textit{bearing-environment} \\ k_l \frac{\partial T_T}{\partial z} = k_b \frac{\partial T_T}{\partial z} & \textit{bearing-lubricant} \\ k_J \frac{\partial T_T}{\partial z} = \frac{1}{2\pi} k_l \int_0^{2\pi} \frac{\partial T_T}{\partial z} \mathrm{d}\theta & \textit{journal-lubricant} \end{cases} \tag{3.10}$$

In this work, based on the continuous sliding characteristics of the lubricant between the two parts, equations were established at the common interface where the pressure, flow, and heat flux in each part are equal. These equations were incorporated into the lubrication calculations to account for the coupled lubrication effects. Specifically, at the common interface, the pressure in the journal component is equal to the pressure in the thrust component:

$$\boldsymbol{P}_J = \boldsymbol{P}_T \tag{3.11}$$

The flow in the journal part is equal to the flow in the thrust part:

$$\left|\frac{h_J^3}{12\eta}\frac{\partial \boldsymbol{P}_J}{\partial z}\right|_{j,m} = \left|\frac{h_T^3}{12\eta}\frac{\partial \boldsymbol{P}_T}{\partial r}\right|_{j,m} \tag{3.12}$$

where the left formula is the axial flow of the journal part, and the right formula is the radial flow of the thrust part. Since the flow is directional, the absolute value is used to represent the flow magnitude. The specific value is determined according to the positive direction of the coordinate system. The heat flow in the journal part is equal to the heat flow in the thrust part:

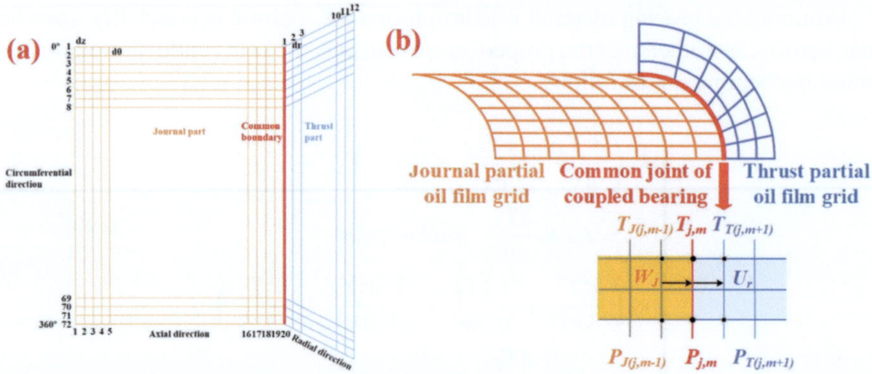

Fig. 3.2 Schematic diagram of flow, pressure and thermal continuity conditions: **a** two-dimensional grid; **b** three-dimensional grid

$$\frac{\left|\rho C_p w_J \frac{\partial T_J}{\partial z} + \lambda \frac{\partial^2 T_J}{\partial z^2}\right|}{Journal\ part} = \frac{\left|\rho C_p U_r \frac{\partial T_T}{\partial r} + \lambda \frac{\partial^2 T_T}{\partial r^2}\right|}{Thrust\ part}$$
$$\left| w_J \frac{\partial T_J}{\partial z} \right|_{j,m} = \left| U_r \frac{\partial T_T}{\partial r} \right|_{j,m} \tag{3.13}$$

Based on the above equations, the pressure, flow, and thermal continuity conditions at the journal-thrust interface of the JTC bearing are considered, thereby revealing the journal-thrust coupled lubrication mechanism of the JTC bearing. Schematic diagram is shown in Fig. 3.2.

As mentioned above, the movement trajectory of the thrust part was directly controlled by the axial shifting displacement. While for the journal part, the transient axis orbit was calculated through the motion equation. By establishing the force balance equation in each direction, that is, $W + Load = ma = m\frac{\mathrm{d}v}{\mathrm{d}t} = m\frac{\mathrm{d}^2e}{\mathrm{d}t^2}$, the final equation of motion is as follows:

$$\begin{cases} e_x^{t+1} = \Delta t^2 \cdot \left(\dfrac{W_x^t - Load_x^t}{m} - \dfrac{e_x^{t-1}}{\Delta t^2} + \dfrac{2e_x^t}{\Delta t^2} \right) \\ e_y^{t+1} = \Delta t^2 \cdot \left(\dfrac{W_y^t - Load_y^t}{m} - \dfrac{e_y^{t-1}}{\Delta t^2} + \dfrac{2e_y^t}{\Delta t^2} \right) \end{cases} \tag{3.14}$$

Before conducting the simulations, grid independence was verified, and the final grid discretization chosen was 72 × 20 × 4 (circumferential × axial × radial) for the journal part and 72 × 4 × 12 for the thrust part, and the hydrodynamic mesh structure of the model can be found in Fig. 3.3. For the engine model in this study, a full working cycle is 720 degrees of crankshaft rotation. Using a time step of 0.05 degrees of crankshaft rotation, the time step Δt was calculated as $\Delta t = 60/\omega/720 * 0.05$.

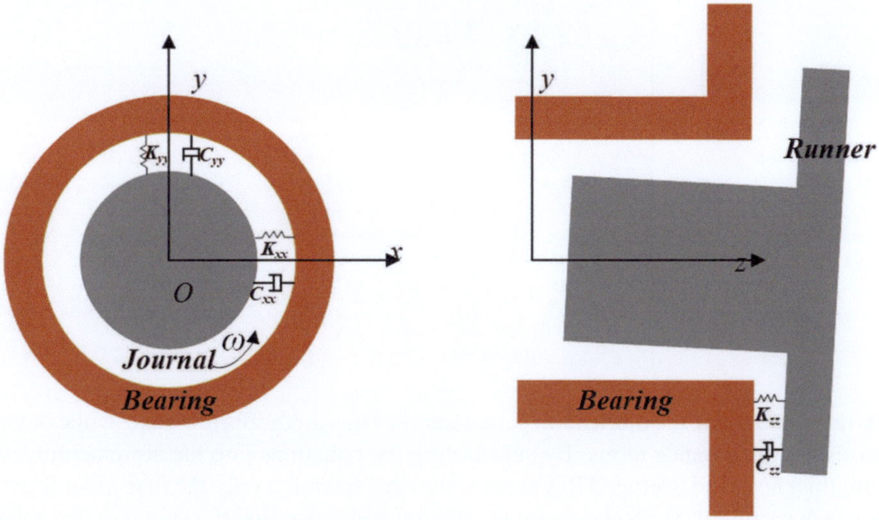

Fig. 3.3 Schematic diagram of the dynamic coefficients

3.2.2 *Dynamic Coefficient Model*

Dynamic coefficients describe the response and behavior of the system in a dynamic environment through stiffness and damping, which is shown in Fig. 3.3. To predict the natural characteristics of the crankshaft system, bearing stiffness calculation are required.

The JTC bearing is located at the end of the crankshaft and is subject to complex forces. Under time-varying multi-directional excitations, axial movement is prone to occur, which can lead to bearing wear and failure. Therefore, it is necessary to accurately assess its dynamic characteristics. This work establishes a dynamic characteristic calculation model that considers horizontal, vertical, and axial directions based on the structure and operating conditions of the JTC bearing. It also conducts time-varying dynamic parameter calculations on the basis of an already well-considered lubrication model, enabling real-time updates of the bearing's time-varying dynamic parameters. Essentially, by applying a small perturbation method to the steady-state Reynolds equation with small perturbation velocities and displacements, disturbance Reynolds equations in the three directions are obtained. Expanding the film thickness, its time derivative, and pressure yields the following form (Cao et al. 2022):

$$
\begin{cases}
h_{J/T} = h_0 + \underbrace{\sum \dfrac{\partial h_{J/T}}{\partial \zeta} \Delta \xi}_{\text{Displacement disturbance}} \\
\dfrac{\partial h}{\partial t} = \dfrac{\partial h_0}{\partial t} + \underbrace{\sum \dfrac{\partial h}{\partial \zeta} \Delta \dot{\xi}}_{\text{Velocity disturbance}} + \underbrace{\sum \dfrac{\partial}{\partial t}\left(\dfrac{\partial h}{\partial \zeta}\right) \Delta \xi}_{\text{Displacement disturbance}} \\
P_{J/T} = P_0 + \underbrace{\sum \dfrac{\partial P_{J/T}}{\partial \zeta} \Delta \xi}_{\text{Displacement disturbance}} + \underbrace{\sum \dfrac{\partial P_{J/T}}{\partial \dot{\zeta}} \Delta \dot{\xi}}_{\text{Velocity disturbance}}
\end{cases}
\tag{3.15}
$$

Here, h_0 and p_0 are the lubrication parameters in a quasi-equilibrium state, while other terms are disturbance terms. By substituting the equations into the aforementioned film thickness and averaged Reynolds equations, retaining only the first-order terms for each perturbation displacement or velocity and separating the variables, the final equations are as follows:

$$
\begin{aligned}
&\frac{\partial}{R\partial\theta}\left(\phi_\theta \frac{h_J^3}{12\eta}\frac{\partial p_{J_\zeta}}{R\partial\theta}\right) + \frac{\partial}{\partial z}\left(\phi_z \frac{h_J^3}{12\eta}\frac{\partial p_{J_\zeta}}{\partial z}\right) \\
&= \begin{cases}
-\frac{\partial}{R\partial\theta}\left(\phi_\theta \frac{h_J^2}{4\eta}\frac{\partial p_{J_\zeta}}{R\partial\theta}\sin\theta\right) - \frac{\partial}{\partial z}\left(\phi_z \frac{h_J^2}{4\eta}\frac{\partial p_{J_\zeta}}{\partial z}\sin\theta\right) - \frac{\omega}{2} sin\theta & : J_\zeta = x \\
-\frac{\partial}{R\partial\theta}\left(\phi_\theta \frac{h_J^2}{4\eta}\frac{\partial p_{J_\zeta}}{R\partial\theta}\cos\theta\right) - \frac{\partial}{\partial z}\left(\phi_z \frac{h_J^2}{4\mu}\frac{\partial p_{J_\zeta}}{\partial z}\cos\theta\right) + \frac{\omega}{2} cos\theta & : J_\zeta = y \\
0 & : J_\zeta = z \\
\cos\theta & : J_\zeta = \dot{x} \\
\sin\theta & : J_\zeta = \dot{y} \\
0 & : J_\zeta = \dot{z}
\end{cases}
\end{aligned}
\tag{3.16}
$$

$$
\begin{aligned}
&\frac{\partial}{r\partial\theta}\left(\phi_\theta \frac{h_T^3}{12\eta}\frac{\partial p_{T_\zeta}}{r\partial\theta}\right) + \frac{\partial}{\partial r}\left(\phi_r \frac{h_T^3}{12\eta}\frac{\partial p_{T_\zeta}}{\partial r}\right) \\
&= \begin{cases}
0 & : T_\zeta = x \\
0 & : T_\zeta = y \\
\frac{\partial}{r\partial r}\left(\phi_\theta \frac{r h_T^2}{4\eta}\frac{\partial p_0}{\partial r}\right) + \frac{\partial}{r\partial\theta}\left(\phi_r \frac{h_T^2}{4\eta}\frac{\partial p_0}{r\partial\theta}\right) & : T_\zeta = z \\
0 & : T_\zeta = \dot{x} \\
0 & : \xi_T = \dot{y} \\
1 & : \xi_T = \dot{z}
\end{cases}
\end{aligned}
\tag{3.17}
$$

In the equations, the subscript J_ζ represents the disturbance terms in the journal part, and T_ζ represents the disturbance terms in the thrust part. The disturbances in horizontal, vertical, and axial displacements correspond to ξ being x, y, and z respectively. Similarly, the disturbances in horizontal, vertical, and axial velocities correspond to ξ being $\dot{x}$, $\dot{y}$, and $\dot{z}$ respectively.

Upon completing the lubrication calculations and obtaining the lubrication performance for the current time step, the film thickness and film pressure at this moment were considered to be in quasi-equilibrium state. They were substituted into the perturbed Reynolds equation and solved using the upwind difference method to obtain the perturbed film pressure, with a convergence criterion of 1×10^{-5}. By integrating the obtained perturbed pressure, the stiffness and damping for each part can be calculated, as shown in the following equations:

$$\boldsymbol{K}_J = \iint\limits_J \begin{Bmatrix} -\cos\theta \\ -\sin\theta \\ 0 \end{Bmatrix} \{p_x p_y p_z\} \mathrm{d}\Omega_J = \begin{bmatrix} Kxx & Kxy & Kxz \\ Kyx & Kyy & Kyz \\ 0 & 0 & 0 \end{bmatrix} \tag{3.18}$$

$$\boldsymbol{C}_J = \iint\limits_J \begin{Bmatrix} -\cos\theta \\ -\sin\theta \\ 0 \end{Bmatrix} \{p_{\dot{x}} p_{\dot{y}} p_{\dot{z}}\} \mathrm{d}\Omega_J = \begin{bmatrix} Cxx & Cxy & Cxz \\ Cyx & Cyy & Cyz \\ 0 & 0 & 0 \end{bmatrix} \tag{3.19}$$

$$\boldsymbol{K}_T = \iint\limits_T \begin{Bmatrix} 0 \\ 0 \\ -1 \end{Bmatrix} \{p_x p_y p_z\} d\Omega_T = \begin{bmatrix} 0 & 0 & 0 \\ 0 & 0 & 0 \\ Kzx & Kzy & Kzz \end{bmatrix} \tag{3.20}$$

$$\boldsymbol{C}_T = \iint\limits_T \begin{Bmatrix} 0 \\ 0 \\ -1 \end{Bmatrix} \{p_{\dot{x}} p_{\dot{y}} p_{\dot{z}}\} \mathrm{d}\Omega_T = \begin{bmatrix} 0 & 0 & 0 \\ 0 & 0 & 0 \\ Czx & Czy & Czz \end{bmatrix} \tag{3.21}$$

After calculating the dynamic characteristics of the bearing at this time step, the eccentricity and the geometric film thickness in thrust part for the next time step were computed based on Eqs. (3.14) and (3.2). Subsequently, the next time step calculation was carried out until the working cycle was complete. Both the lubrication model and the dynamic characteristics model in this work were programmed and calculated using Fortran. Unlike conventional calculations where stiffness and damping were set as constants, the dynamic characteristic parameters obtained in this work were calculated based on a more accurate lubrication model and can reflect the transient dynamic performance parameters under time-varying loads. This allows for the diagnosis of bearing failure causes under sudden load changes. The specific calculation flowchart can be found in Fig. 3.4.

3.2.3 Model Verification

For the verification of the lubrication model, a JTC bearing test rig was built and the experiment with dynamic load was carried out. Figure 3.5 shows a photo and schematic of the test rig. Considering that the impact of the crankshaft system environment on JTC bearing lubrication mainly manifests in time-varying multi-directional loads. Therefore, although the entire crankshaft was not fully built in the

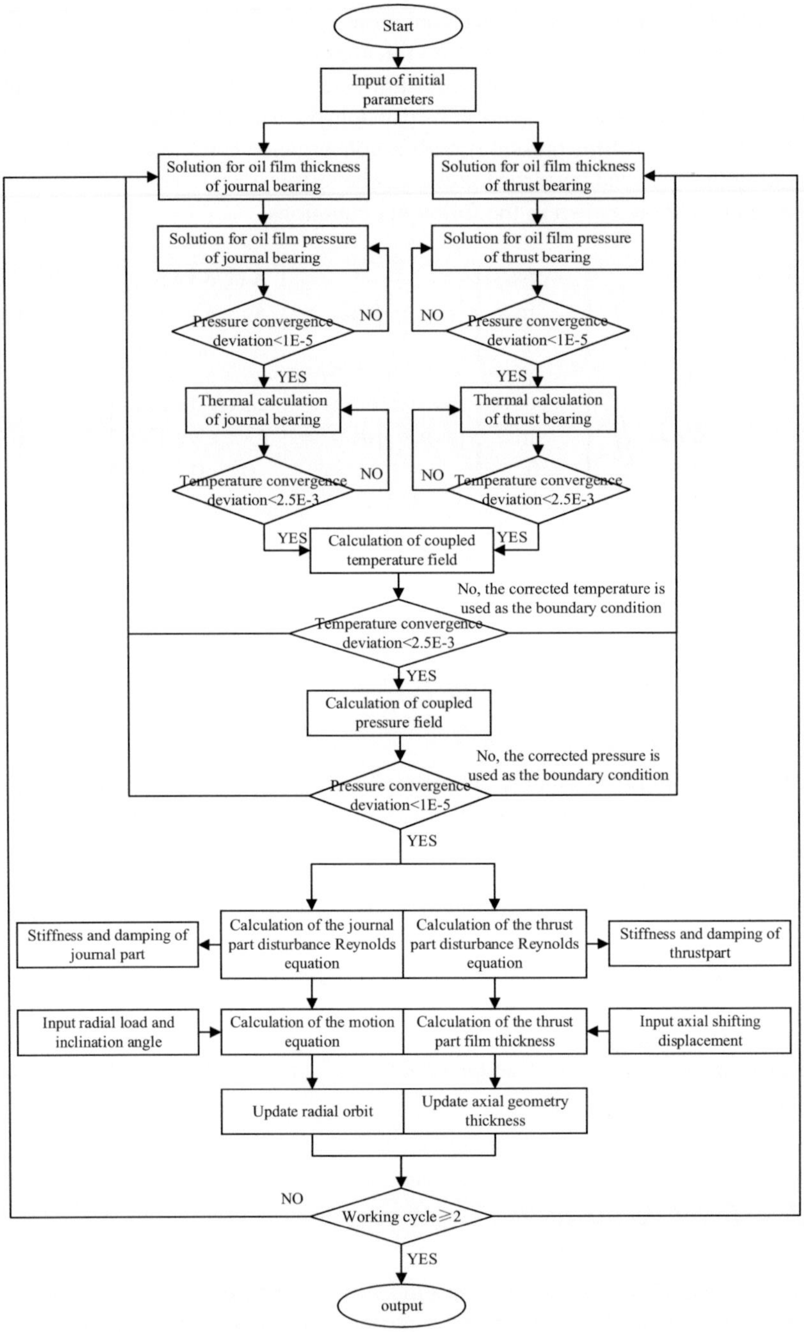

Fig. 3.4 Flow chart of the calculation

test rig, the actual crankshaft system working environment was approximately simulated by directly applying dynamic sinusoidal loads and axial forces to the tested bearing. A continuous sinusoidal load was applied to the bearing under test through the radial loading device at the bottom, a constant axial force was applied through the axial loading device on the left, and oil was continuously supplied through the oil box above. Based on the torque sensor, the torque of the bearing was obtained and the friction coefficient was calculated. The resistance temperature sensor (PT100) was used to obtain the bearing surface temperature. Table 3.1 shows relevant experimental parameters, and the radial load used for the test can be found in Fig. 3.6a. Negative values represent the load direction being vertically downward. The test was run for an hour to ensure the results were stable. Figure 3.6b illustrates the comparison of friction coefficient, and the result calculated by the original model was also added for comparison. The calculation model of the friction coefficient is as follows:

$$\mu = \frac{R_1 \int\int \left(\frac{\eta U_J}{h_J} + \frac{h_J}{2}\frac{\partial P_J}{\partial \theta}\right) R_0 \mathrm{d}\theta \mathrm{d}z + \frac{(R_1+R_2)}{2} \iint \left(\frac{\eta U_J}{h_T} + \frac{h_T}{2}\frac{\partial P_T}{\partial \theta}\right) r \mathrm{d}\theta \mathrm{d}r}{\int\int P_J R_1 \mathrm{d}\theta \mathrm{d}z} \tag{3.22}$$

The friction coefficient calculation method of the proposed model and the original model was the same, while the calculated film pressure was different, resulting in different calculated friction coefficient results. It can be found that the trend is consistent, and the calculation results of the proposed model are closer to the experimental

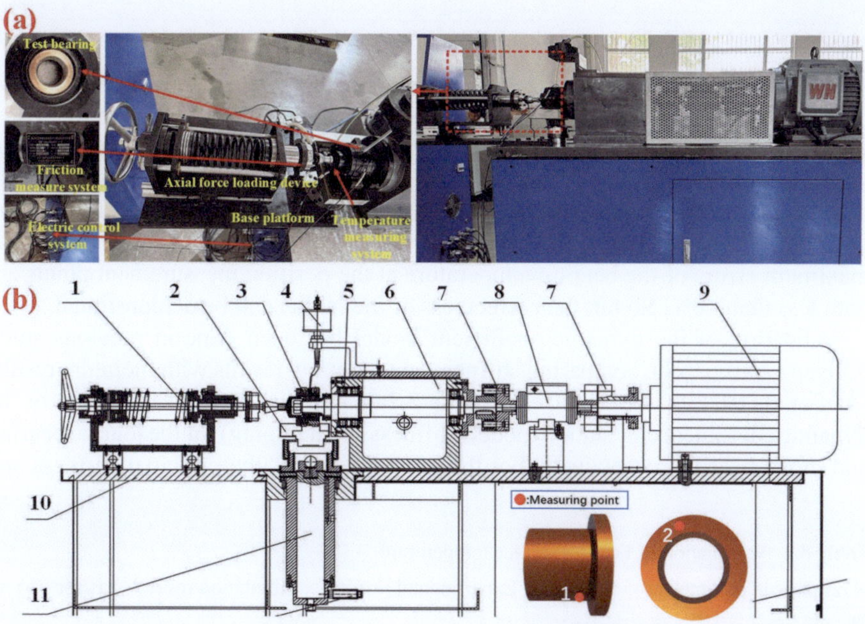

Fig. 3.5 A photo and schematic of the test rig: **a** photo of the test rig; **b** schematic of the test rig

Table 3.1 Experimental parameters

Parameter	Value
Inner radius of journal part	20 mm
Outer radius of thrust part	30 mm
Inclination angle	0.01°
Angular velocity	500 rpm
Viscosity	0.113 Pa s
Supply temperature	35 °C
Horizontal load	300 N

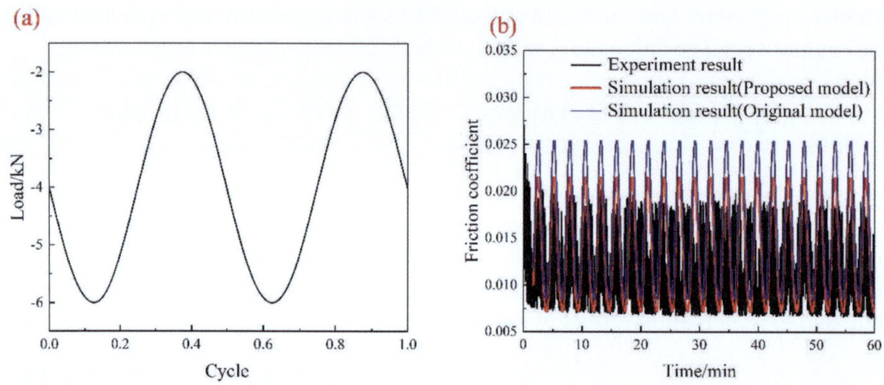

Fig. 3.6 Radial load for the test and the verification result of friction coefficient: **a** experimental radial load; **b** verification results of friction coefficient

values, with the maximum amplitude deviation in the entire test being approximately 6%. Table 3.2 shows the comparison of the bearing surface temperature at two measuring points. The experimental result was the average temperature of the last 10 min, and the simulation result was the calculation result of one cycle. The maximum errors of the bearing temperature at the position measurement points are both less than 10%. So far, the correctness of the model can be demonstrated.

Validation of the dynamic coefficient model has been done in previous study (Oliveira et al. 2021), comparing stiffness and damping results with literature results (Arif et al. 2022; Xie et al. 2021). Figure 3.7a and b display the verification of the dynamic coefficient calculation model (stiffness and damping) for the journal bearing and thrust bearing respectively. For the journal bearing, the principal stiffness and

Table 3.2 Verification of bearing surface temperature

Comparison parameters	Experimental result	Simulation result	Error (%)
Temperature at measuring point 1	43.3	45.8	5.7
Temperature at measuring point 2	46.3	48.5	4.8

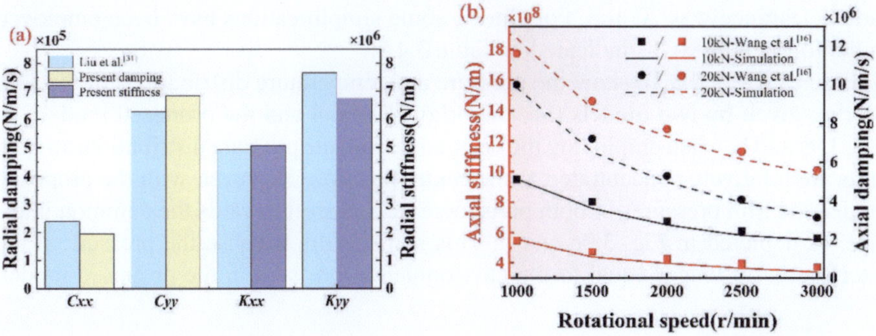

Fig. 3.7 Verification of dynamic coefficient model: **a** verification of journal bearing; **b** verification of thrust bearing

damping (horizontal and vertical) at fixed eccentricity and tilt angles were calculated based on the parameters from reference (Xie et al. 2021). It was found that the simulation results were in good agreement with the literature values, with an average deviation of less than 10% and a maximum deviation of less than 20%. Similarly, for the thrust bearing, the axial stiffness and damping variations with rotational speed under two load conditions were calculated based on reference (Arif et al. 2022). The simulation results showed trends consistent with the literature, with an average deviation of less than 10% and a maximum error of less than 11%. This confirms the accuracy of the dynamic coefficient model.

3.3 Analysis of the Transient Lubrication and Dynamic Coefficient of Journal-Thrust Coupled Bearing Under Simple Displacement

3.3.1 Necessity Analysis

The environmental and structural parameters utilized in this study are outlined in Table 3.3. To more accurately simulate real working conditions, the radial load was computed based on the actual cylinder explosion pressure utilizing the multi-body dynamics software AVL EXCITE Power Unit (Ma et al. 2003). The cylinder pressure curve is depicted in Fig. 3.8. The axial load is directly determined by the axial shifting Eq. (3.3) to replicate axial shifting movement. To investigate the discrepancies in lubrication results calculated by different models, a comparison was conducted between the results considering the coupled effects and those without considering coupled effects. The distribution results at a specific time (T/6) within the cycle are displayed in Figs. 3.9 and 3.10. For this section, the incline angle was set to 0.001°, the inlet oil temperature was 40°C, the geometry film thickness was 22μm, and the radial clearance was 200μm. For clarity, some simplifications have been employed in the following text, as indicated in Table 3.4.

Figures 3.9 and 3.10 show the pressure and temperature distributions of the JTC bearing given by two models (i.e. the original model and the proposed model). In Fig. 3.9a and b, when employing the original model, the pressure distributions in both parts are relatively concentrated at the center position. However, with the proposed model, the film pressures of both parts increase and shift towards the common interface, as depicted in Fig. 3.9c and d. This shift occurs because the pressure at the interface is no longer equal to the environmental pressure in the proposed model.

Table 3.3 Environmental and structural parameters for calculation

Parameter	Value
Radius clearance/c	100/200/300/400 μm
Incline angle of journal/γ	0.0001/0.0005/0.0007/0.001°
Supply temperature/T_0	30/40/50/60 °C
Brand of the oil/η_0	15W40
Elastic modulus/E_J	105 GPa
Poisson ratio/v_p	0.327
Velocity/ω	36.65 rad/s
Viscosity temperature coefficient/β	− 0.0491
Thermal expansion coefficient/k_p	1.2E−5/K
Thermal conductivity/k_b	47 W/mK
Heat transfer coefficient/h_b	80 W/m^{2}K
Thermal conductivity/k_r	0.13 W/mK
Specific heat capacity/Cp_o	2000 W/m^{2}K

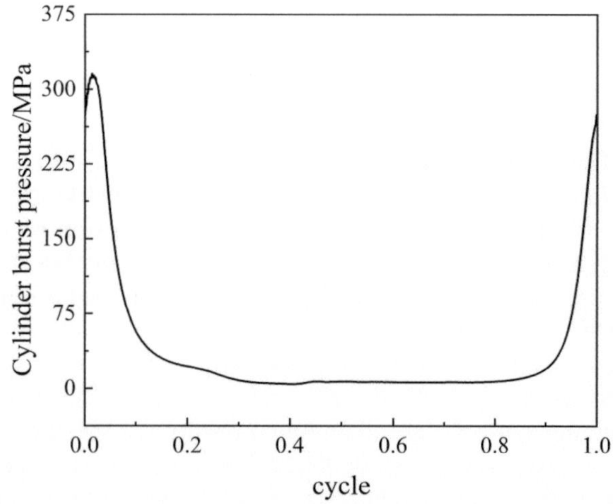

Fig. 3.8 Cylinder explosion pressure of the focused engine

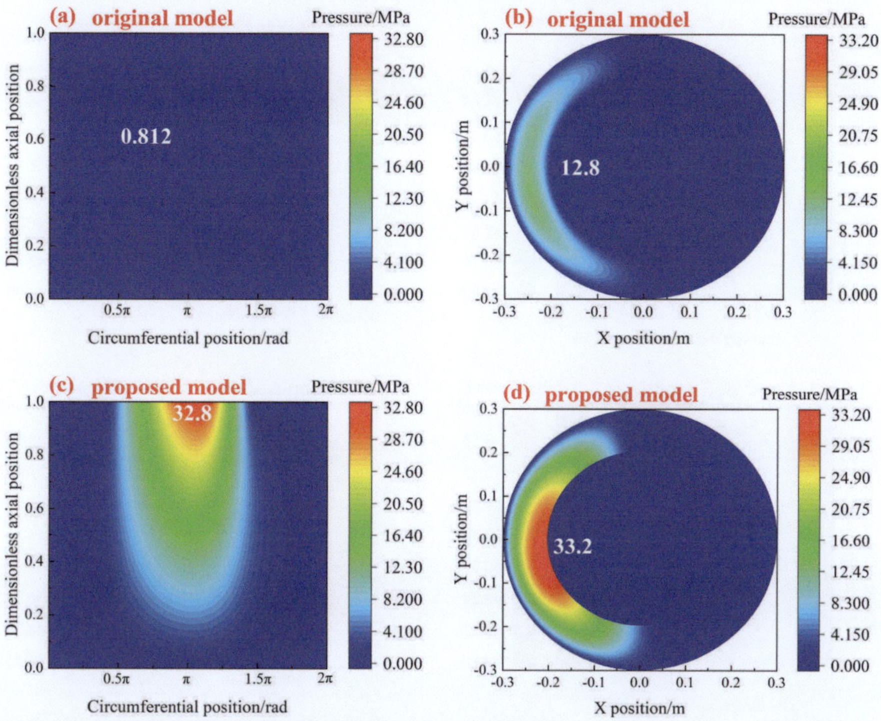

Fig. 3.9 Film pressure distribution of each part under different models at T/6: **a** journal part (calculated by original model); **b** thrust part (calculated by original model); **c** journal part (calculated by proposed model); **d** thrust part (calculated by proposed model)

Influenced by the continuity conditions of flow and pressure, a continuous pressure distribution emerges at the common boundary. Additionally, the maximum pressure (P_{max}) in both parts increases, especially in the thrust part. This increase can be attributed to the rise in pressure at the common interface under the thermal-pressure coupled effect, causing the pressures of each part to mutually influence each other. Given that the original P_{max} value in the thrust part is significantly higher than that in the journal part, the proposed film pressure of the journal part is notably affected, resulting in a larger increase.

Similarly, Fig. 3.10a and b reveal that the results calculated by the original model exhibit smaller temperatures, with the maximum temperature (T_{max}) also appearing at the center of both parts. In contrast, as shown in Fig. 3.10c and d, T_{max} of both parts increases, particularly for the thrust part, when using the proposed model. This temperature increase is attributed to the proposed model raising the inlet oil temperature of the thrust part from ambient temperature to the outlet oil temperature of the journal part. Consequently, there is an overall elevation in the film temperature of the thrust part. Given that temperature is significantly influenced by pressure, the distribution of temperature aligns with the pressure distribution.

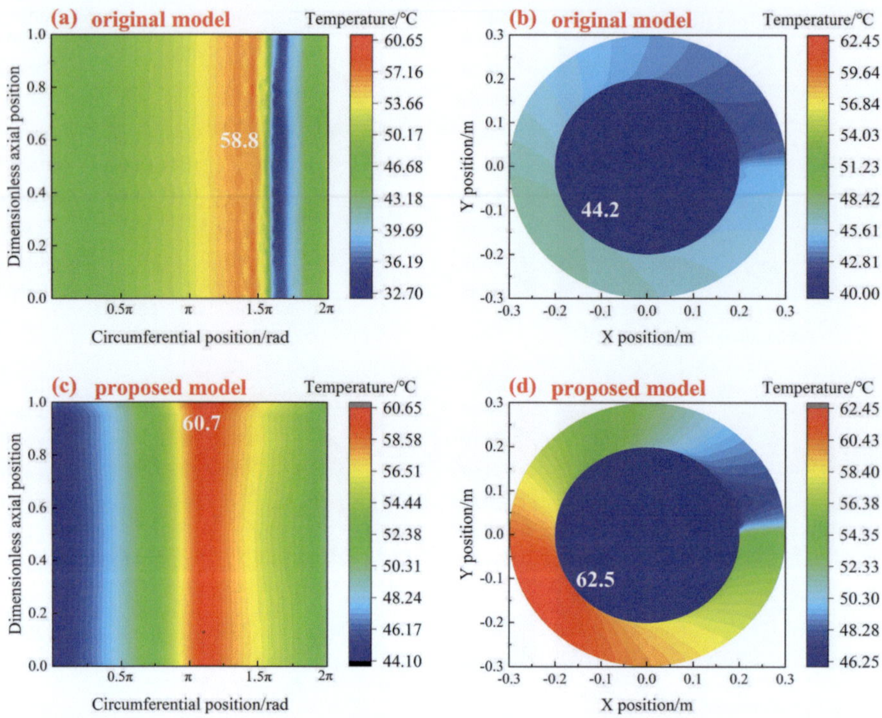

Fig. 3.10 Film temperature distribution of each part under different models at T/6: **a** journal part (calculated by original model); **b** thrust part (calculated by original model); **c** journal part (calculated by proposed model); **d** thrust part (calculated by proposed model)

Table 3.4 Explanation of abbreviation

Abbreviation	Explanation
Thermal-pressure coupled effect	Coupled effects on lubrication caused by flow, pressure and thermal continuity
Original model	Transient TEHD lubrication model without considering the thermal-pressure coupled effect
Proposed model	Transient TEHD lubrication model considering the thermal-pressure coupled effect
$P_{\max}$	Maximum film pressure
$T_{\max}$	Maximum film temperature
Original $P_{\max}$	$P_{\max}$ calculated by the original model
Original $T_{\max}$	$T_{\max}$ calculated by the original model
Proposed $P_{\max}$	$P_{\max}$ calculated by the proposed model
Proposed $T_{\max}$	$T_{\max}$ calculated by the proposed model

3.3.2 Influences of Transient Coupled Effects Under Various Axial Clearances

To elaborate on the effects of the thermal-pressure coupled effect on lubrication performance, a parameterized analysis was conducted with variations in different input parameters. This analysis primarily focuses on two aspects: the journal part and the thrust part. The input parameters considered for the journal part include waveforms of radial load, incline angles, inlet oil temperatures, and radial clearances. In contrast, for the thrust part, the key input parameter is axial clearances.

This section investigates the influence of the coupled effect on lubrication characteristics under four axial clearances. The geometry film thickness (h_0) in Eq. (3.2) served as the control variable affecting the axial clearance, with lubrication calculations conducted for h_O values of 22 μm, 32 μm, 42 μm, and 52 μm, respectively. Figures 3.11 and 3.12 illustrate the lubrication characteristics of the two models under four types of axial clearances, and Fig. 3.13 displays proposed P_{max} and proposed T_{max} under four axial clearances at T/6. In Fig. 3.11a, the proposed P_{max} values for the journal part and thrust part exhibit a significant increase over a working cycle compared to the original results. Additionally, the shapes of P_{max} calculated under the two models differ. Specifically, at the beginning and end stages of the cycle, the curve of the proposed P_{max} in the journal part changes from a smooth straight line to a wave curve compared with the original results. This is attributed to the fact that when using the 'TEHD coupled model,' the pressure in each part at the common interface increases and begins to influence each other. Given that the original P_{max} value in the thrust part is much higher than that in the journal part, the proposed P_{max} value in the journal part increases significantly and follows the same trend as in the thrust part. The specific distributions of the proposed film pressure in each part at time T/6 are shown in Fig. 3.11b, and proposed P_{max} is illustrated in Fig. 3.13a. As h_0 increases from 22 to 52 μm, P_{max} in each part gradually decreases. In these cases, since the thrust part remains in a state of hydrodynamic lubrication, P_{max} changes almost linearly with the axial clearance. Additionally, P_{max} of the journal and thrust parts are found to be close to each other. This is because the original pressure in the journal part at T/6 is relatively small, not in the same order of magnitude as the pressure in the thrust part. P_{max} of the journal part is dominated by the pressure of the thrust when using the proposed model. It can be inferred that in the actual operation of coupled bearings, if one part experiences a high-pressure load, the lubrication situation of another part will also worsen.

The variation of T_{max} follows a pattern similar to that of P_{max}, as shown in Fig. 3.12a. Compared with the original T_{max}, the proposed T_{max} for each part increases, and the axial clearance begins to exhibit an impact on the journal part. For the journal part, the curve of the proposed T_{max} shows an increase in the initial and final regions of the cycle. In the thrust part, the curve of the proposed T_{max} is similar in shape to the journal part but numerically larger. This is because the oil in the thrust part is supplied by the end leakage of the journal part, and the inlet oil temperature is significantly influenced by the journal part. The specific distributions

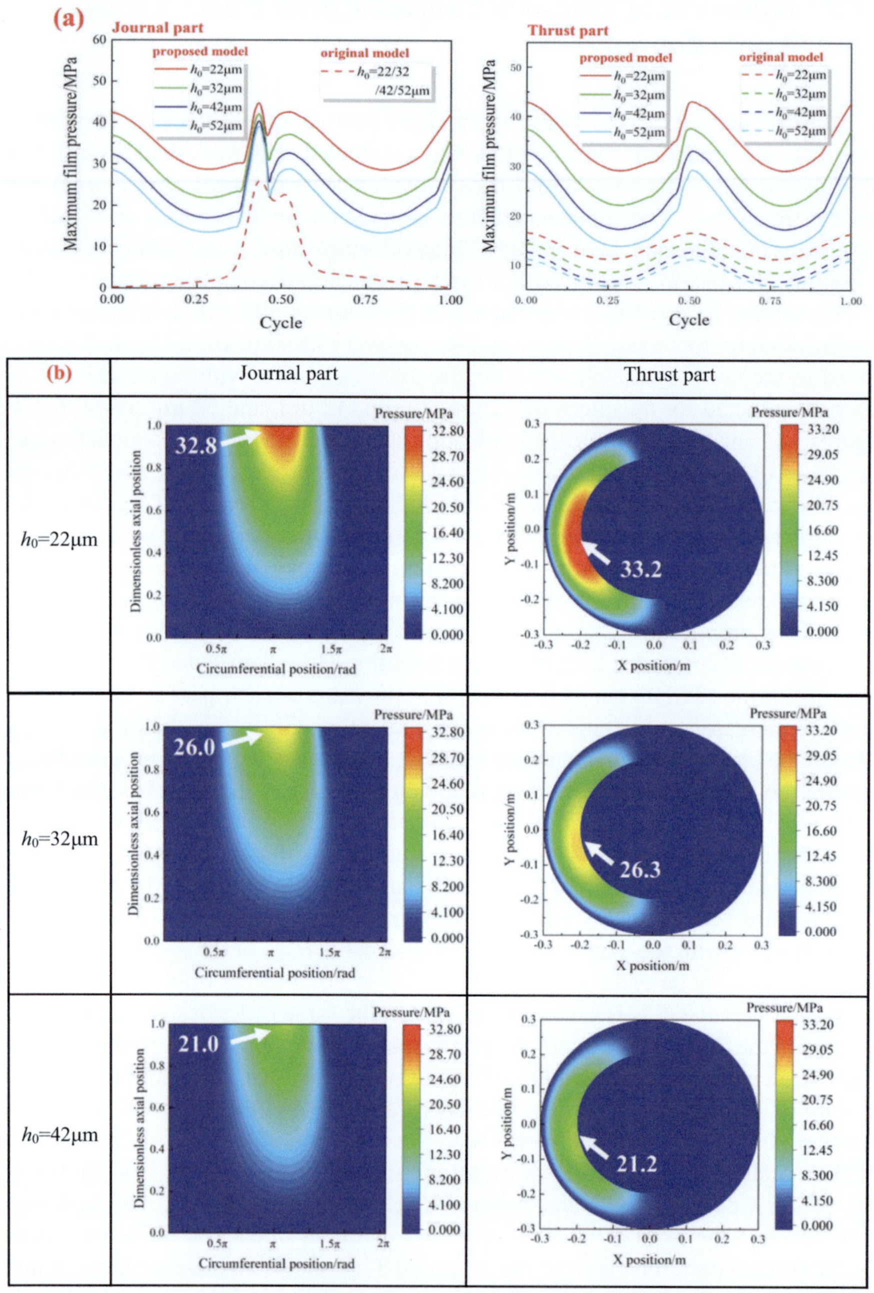

Fig. 3.11 P_{max} of each part under different axial clearances: **a** proposed P_{max} and original P_{max} within a working cycle; **b** proposed film pressure distribution at T/6

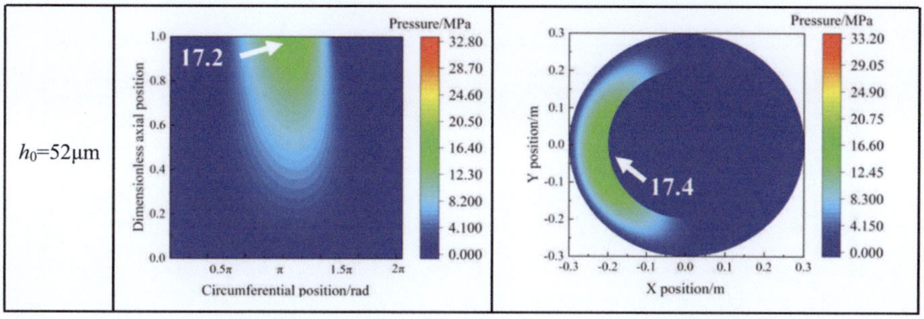

Fig. 3.11 (continued)

of the proposed film temperature in each part at time T/6 are shown in Fig. 3.12b, and the proposed T_{max} is illustrated in Fig. 3.13b. As h_0 increases from 22 to 52 μm, the proposed T_{max} in each part decreases relatively smoothly. It can be observed that the influence of the axial clearance on the temperature is more pronounced under low axial clearances. Based on the above research, it can be inferred that the coupled effect is most pronounced at low axial clearances.

3.3.3 Influences of Transient Coupled Effects Under Various Waveforms of Radial Load

The influences of the coupled effect on lubrication characteristics under four waveforms of radial load were investigated in this section. The lubrication calculation was carried out based on four radial loads with different phases, as shown in Fig. 3.14. To better distinguish the four waveforms of radial load, 'Waveform1', 'Waveform2', 'Waveform3', and 'Waveform4' are used to represent the loads under the four phases, respectively, in the following analysis. Figures 3.15 and 3.16 show the lubrication characteristics of the two models under four waveforms of radial load. As shown in Fig. 3.15a, for the journal part, compared with the original P_{max}, the proposed P_{max} has an increased amplitude, and the trend merges with the curve of P_{max} in the thrust part. However, during a working cycle, the peak position of P_{max} changes with the variation of the radial load. The proposed P_{max} under the radial load corresponding to 'Waveform1' is found to be the highest within a working cycle. This is because at this moment, the loads on the journal and thrust parts are both very high, leading to the greater coupled pressure. For the thrust part, the overall trend does not change much after considering the coupled effect, while a slight decrease occurs at a specific moment. This is caused by the sudden increase in the load of the journal part, which then causes the rapid decrease of the film thickness in the journal part. However, the change rate for the film thickness in the thrust part is slow, leading to a slight decrease in the pressure of the thrust part. The specific distributions of the proposed film pressure in each part at time T/6 are shown in Fig. 3.15b. It can be found that

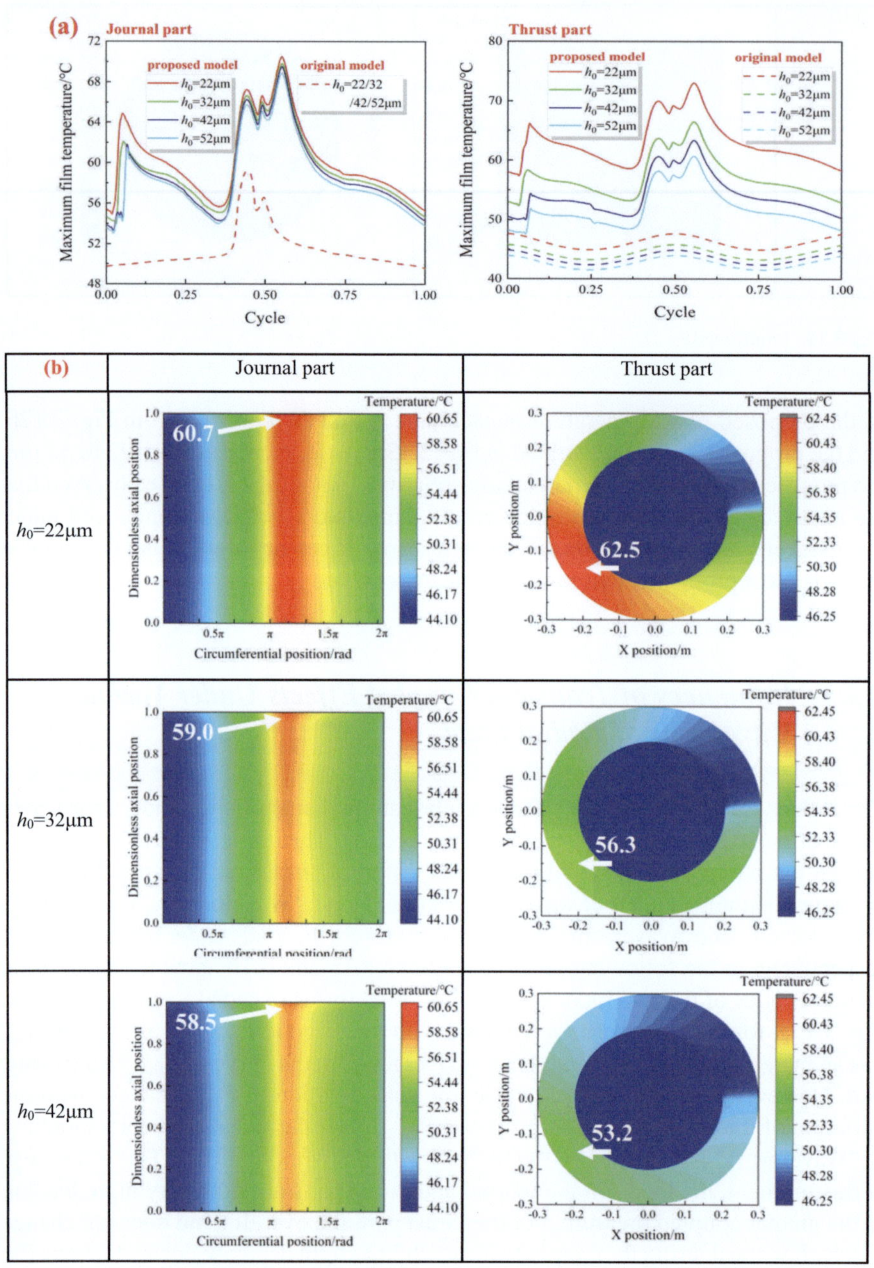

Fig. 3.12 $T_{\max}$ of each part under different axial clearances: **a** proposed $T_{\max}$ and original $T_{\max}$ within a working cycle; **b** proposed film temperature distribution at T/6

h_0=52μm	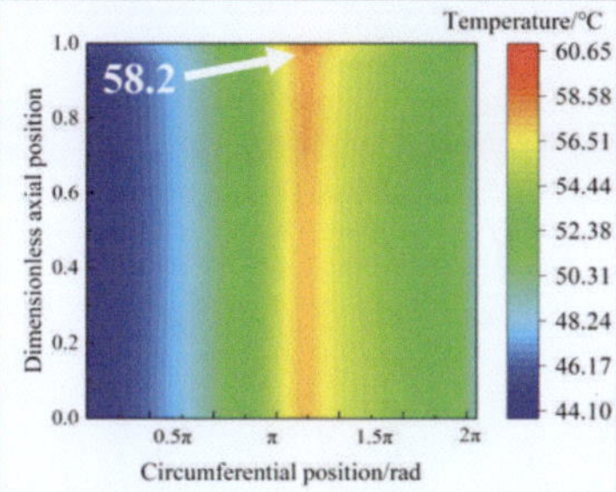	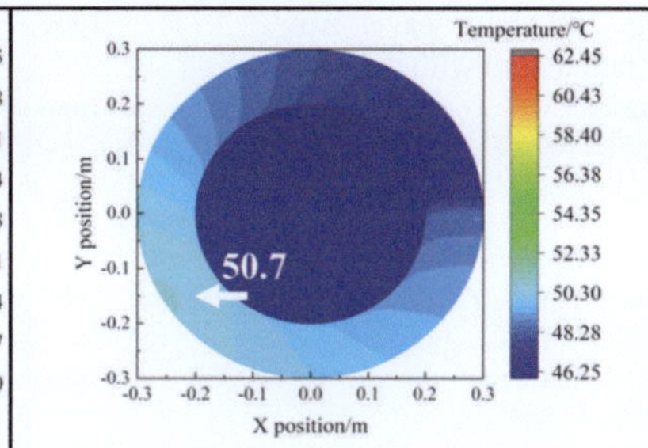

Fig. 3.12 (continued)

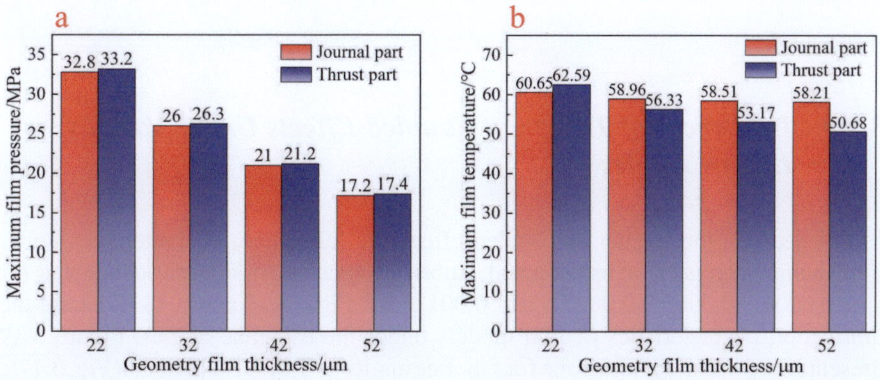

Fig. 3.13 Proposed $P_{\max}$ and $T_{\max}$ under four axial clearances at T/6: **a** proposed $P_{\max}$ in each part; **b** proposed $T_{\max}$ in each part

in four cases, the pressure of each part does not change much, while the radial load corresponding to 'Waveform2' has the highest film pressure in the journal part of 35.1 MPa and the lowest in the thrust part of 31.2 MPa, which corresponds to the analysis results in Fig. 3.15a. It can be inferred that during the working condition design of the JTC bearing, it is advisable to avoid the simultaneous occurrence of high pressure in both parts as much as possible.

Figure 3.16a shows $T_{\max}$ of two models under four radial loads. Similarly, the appearance time of $T_{\max}$ during the working cycle changes consistently with the change of the radial load. The highest $T_{\max}$ of each part is under the radial load corresponding to 'Waveform1', which is consistent with the results above. The specific distributions of the proposed film temperature in each part at time T/6 are shown in Fig. 3.16b. It can be found that the radial load corresponding to 'Waveform4' has the lowest film temperature in each part, which is caused by the phase difference of the radial load.

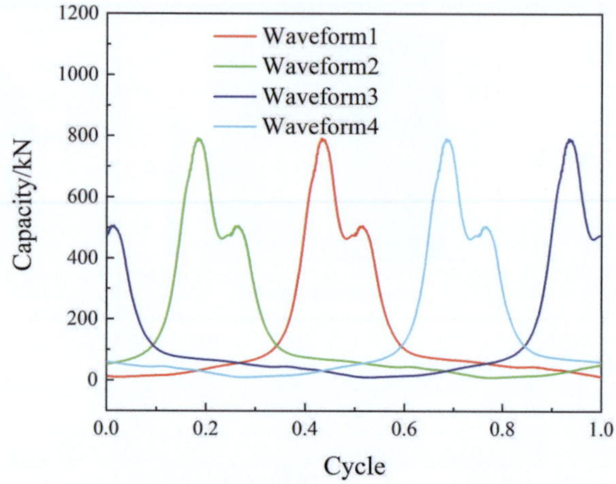

Fig. 3.14 Four waveforms of radial load with different phases

3.3.4 Influences of Transient Coupled Effects Under Various Incline Angles

In this section, the impact of coupled effects on lubrication characteristics under four incline angles (γ) is investigated. Lubrication calculations were conducted for γ of 0.0001°, 0.0005°, 0.0007°, and 0.001°, and Figs. 3.17 and 3.18 illustrate the lubrication characteristics of two models under these incline angles. Figure 3.19 presents P_{max} and T_{max} under the four incline angles at T/6. As depicted in Fig. 3.17a, the curve of the original P_{max} for the journal part remains relatively stable under these incline angles. However, for the thrust part, the original P_{max} noticeably increases with the incline angle, signifying the significant influence of the incline angle on the film pressure in the thrust part. When employing the 'TEHD' model, the shape of P_{max} for the journal part undergoes significant changes, particularly evident in the curves at the beginning and end of the working cycle. Due to the incline angle's increase, P_{max} of the thrust part experiences a substantial rise, exerting a greater influence on the journal part. For the thrust part, when the incline angle is 0.0001°, the coupled effect is not pronounced, resulting in the proposed P_{max} showing marginal increments compared to the original P_{max}, with no significant change in curve shape. However, as the incline angle increases to 0.0005° and 0.0007°, the proposed P_{max} sees a substantial increase due to the thermal-pressure coupled effect, and the curve of P_{max} exhibits fluctuations. Yet, when the incline angle reaches 0.001°, although an overall improvement in P_{max} is observed, the curve returns to a smoother shape. This is attributed to the much higher pressure on the thrust part compared to the journal part, reducing the influence of the journal part on the thrust part. The specific distributions of the proposed film pressure in each part at time T/6 are shown in Fig. 3.17b, and the proposed P_{max} is presented in Fig. 3.19a. As γ increases from 0.0001° to 0.001°, P_{max} in each part undergoes a substantial increase. Moreover, the rise from 0.0001° to 0.0005° results in a 78.9% increase in proposed P_{max} for the

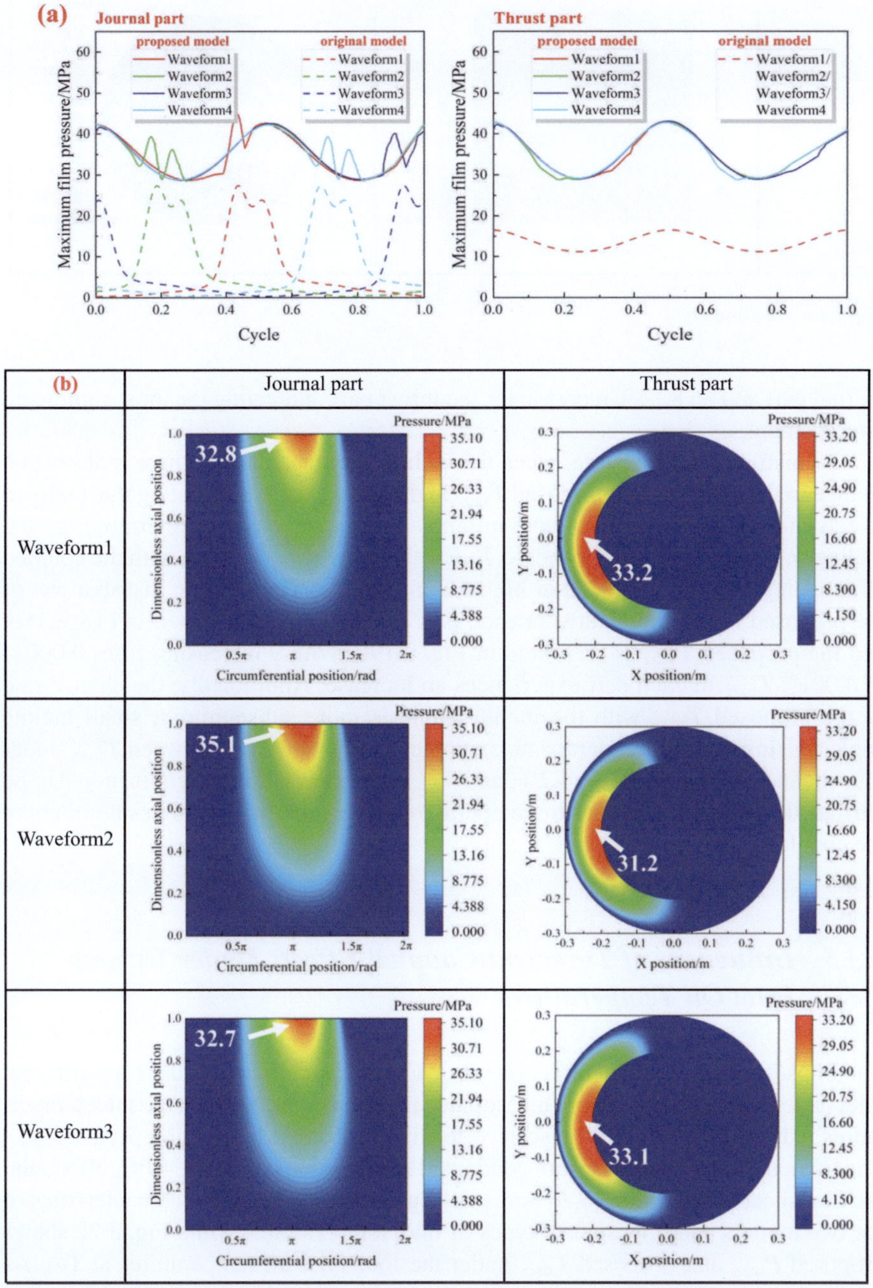

Fig. 3.15 $P_{\max}$ of each part under different waveforms of radial load: **a** proposed $P_{\max}$ and original $P_{\max}$ within a working cycle; **b** proposed film pressure distribution at T/6

Waveform4	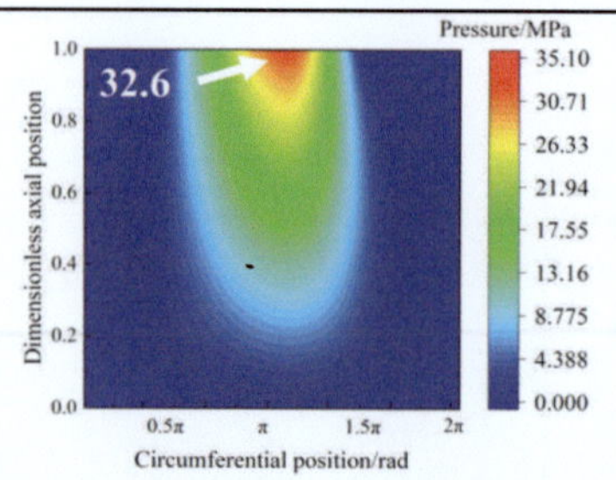	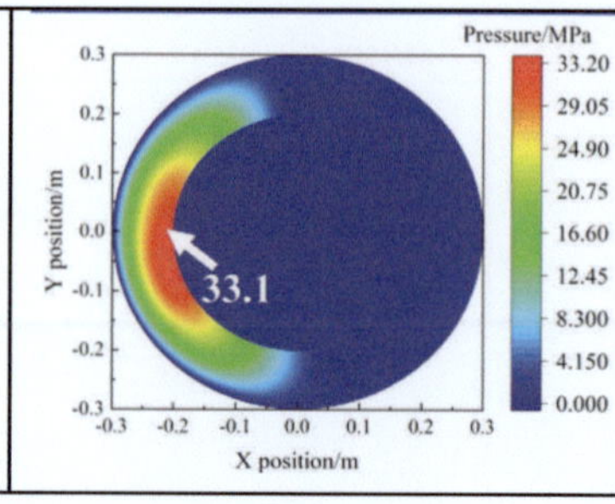

Fig. 3.15 (continued)

journal part and an 82.6% increase for the thrust part, indicating the most significant improvement.

As illustrated in Fig. 3.18a, when the incline angle is 0.0001°, there is almost no change in the shape of the proposed T_{max} in the journal part, suggesting that coupled effects at low incline angles have minimal impact on T_{max}. Furthermore, as the incline angle increases, the proposed T_{max} value for each part rises, with the coupled effect being most pronounced at high incline angles. The specific distributions of the proposed film temperatures in each part at time T/6 are depicted in Fig. 3.18b, and the proposed T_{max} is presented in Fig. 3.19b. With γ increasing from 0.0001° to 0.001°, T_{max} in each part experiences an increase. Additionally, the change rate of the proposed T_{max} with the incline angle is more substantial at small incline angles. Using the journal part as an example, the change rates between T_{max} under four incline angles are 9.5%, 4.2%, and 4%, respectively. This phenomenon can be attributed to the thrust part's film temperature under high incline angles dominating the coupled lubrication.

3.3.5 *Influences of Transient Coupled Effects Under Various Inlet Oil Temperatures*

The impacts of the coupled effect on lubrication characteristics under four different inlet oil temperatures (T_0) were investigated in this section. Varying the inlet temperatures induces changes in oil viscosity, leading to distinct lubrication performance. The lubrication calculations were conducted for T_0 values of 30°C, 40°C, 50°C, and 60°C, respectively. Figures 3.20 and 3.21 illustrate the lubrication characteristics of the two models under these four types of inlet temperatures, while Fig. 3.22 shows proposed P_{max} and proposed T_{max} under the four inlet oil temperatures at T/6. As depicted in Fig. 3.20a, the original P_{max} in the thrust part increases with decreasing inlet oil temperature, while the original P_{max} of the journal part slightly decreases. This behavior is attributed to the decrease in oil viscosity with increasing temperature, resulting in higher film pressure calculated by the Reynolds equation under

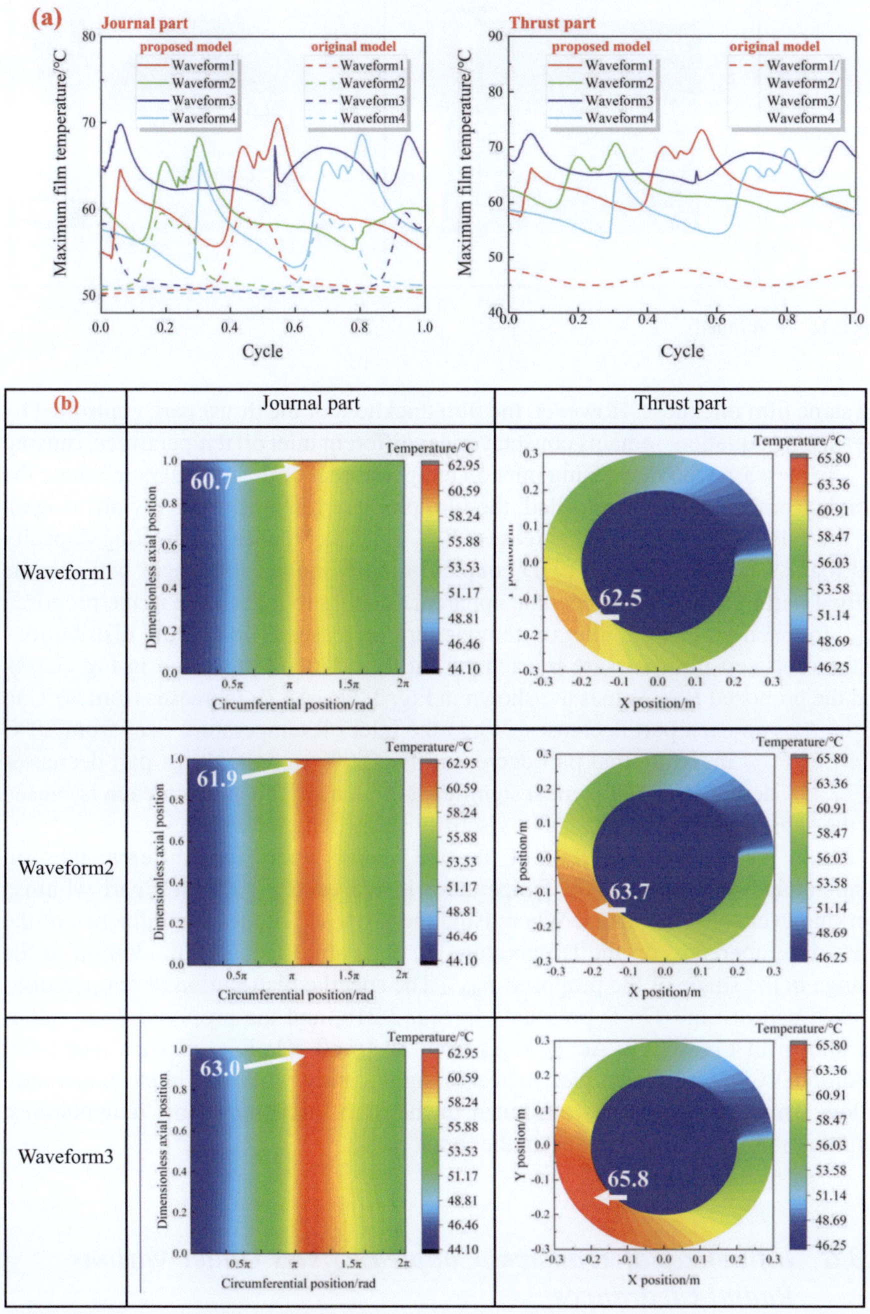

Fig. 3.16 T_{max} of each part under different waveforms of radial load: **a** proposed T_{max} and original T_{max} within a working cycle; **b** proposed film temperature distribution at T/6

Waveform4	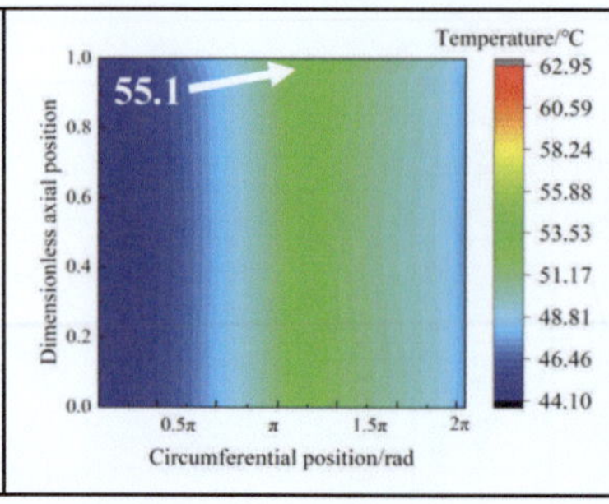	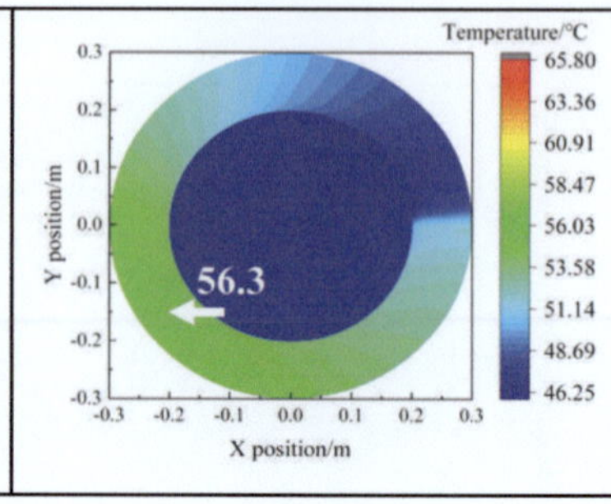

Fig. 3.16 (continued)

the same film thickness. However, the film thickness of the thrust part, controlled by the shifting equation, remains constant under different inlet oil temperatures, causing P_{max} to increase with decreasing inlet oil temperature. For the journal part, where the radial clearance is load-controlled, the film pressure remains relatively unchanged. As the inlet oil temperature increases, P_{max} rises to balance the bearing capacity and load. When using the 'TEHD coupled' model, mutual influences between the journal part and thrust part become apparent, leading to a decrease in the proposed P_{max} for each part as the inlet oil temperature increases. The specific distributions of the proposed film pressure in each part at time T/6 are presented in Fig. 3.20b, and the proposed P_{max} values are shown in Fig. 3.22a. As T_0 increases from 30°C to 60°C, P_{max} in each part decreases. When the inlet oil temperature rises from 30°C to 40°C, P_{max} in the journal part decreases by 32.9%, and the thrust part decreases by 37.1%, demonstrating the most significant influence. This phenomenon is caused by the substantial rate of change in viscosity at low temperatures.

Figure 3.21a illustrates T_{max} of the two models under four different inlet oil temperatures. As the inlet oil temperature increases, T_{max} of each part within a working cycle also increases. When using the proposed model, the influence of the inlet oil temperature on the temperature curve is almost linear, in addition to the change in the shape of the proposed T_{max}. The specific distribution of temperatures in each part at time T/6 is presented in Fig. 3.21b, and the proposed T_{max} values are shown in Fig. 3.22b. As T_0 rises from 30 to 60 °C, T_{max} of each part rises. Additionally, T_{max} in each part increases significantly when the inlet temperature is low. This phenomenon is attributed to the high viscosity at low temperatures, resulting in higher film pressure in the thrust part.

3.3.6 Influences of Transient Coupled Effects Under Various Radial Clearances

The effects of the coupled effect on lubrication characteristics under four different radial clearances (c) were investigated in this section. The lubrication calculation was

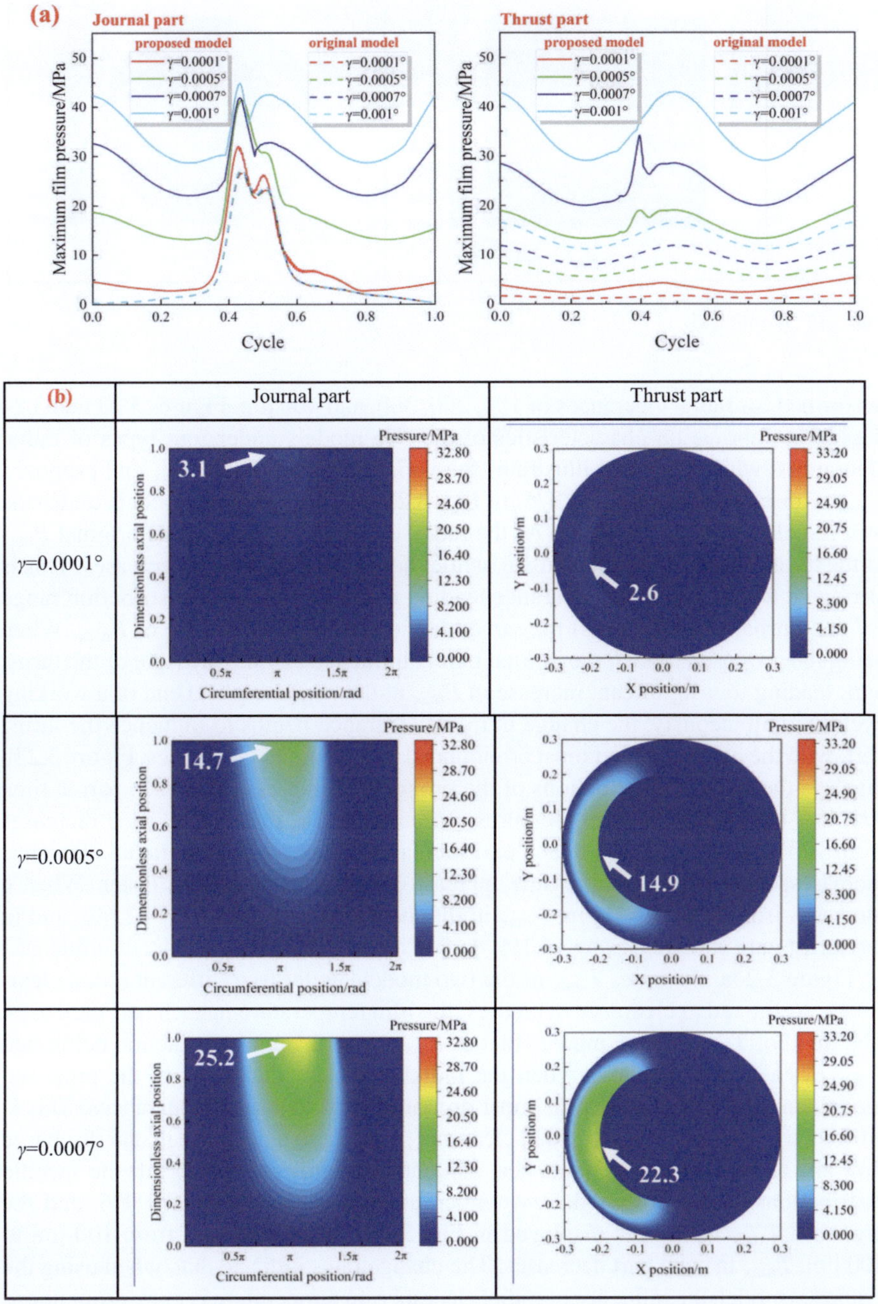

Fig. 3.17 $P_{\max}$ of each part under different incline angles: **a** proposed $P_{\max}$ and original $P_{\max}$ within a working cycle; **b** proposed film pressure distribution at T/6

γ=0.001°

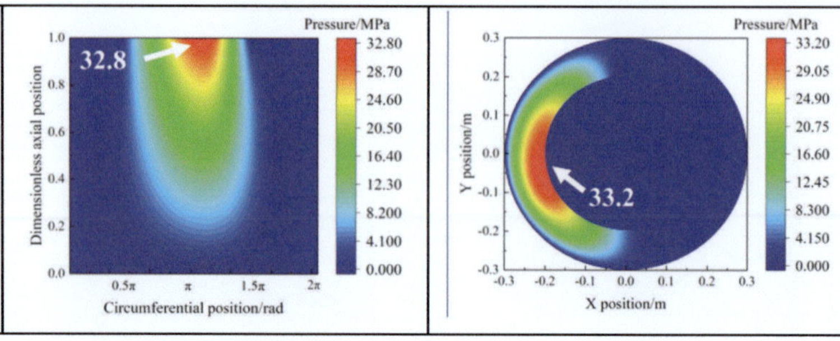

Fig. 3.17 (continued)

performed for radial clearances of 100, 200, 300, and 400 μm. Figures 3.23 and 3.24 depict the lubrication characteristics of the two models under four types of radial clearances, while Fig. 3.25 illustrates the variation of proposed P_{max} and proposed T_{max} with radial clearances at T/6. In Fig. 3.23a, P_{max} of the two models under the four radial clearances is shown. As the radial clearance decreases, the original P_{max} of the thrust part remains unchanged, while that of the journal part decreases. This is attributed to a larger radial clearance leading to a wider pressure distribution range for the journal part to support the same load, resulting in a decrease of P_{max}. When using the proposed model, the journal part is influenced by the pressure of the thrust part, leading to a significant increase in P_{max} at the beginning and end of a working cycle. Simultaneously, the change in radial clearance begins to influence the thrust part, with the coupled effect most pronounced at high radial clearances. Figure 3.23b displays the specific distributions of the proposed film pressure in each part at time T/6, and the maximum film pressures are presented in Fig. 3.25a. As c increases from 100 to 400 μm, P_{max} in each part increases. Moreover, as the radial clearance increases, the rate of film pressure increase in each part becomes faster. When c increases from 300 to 400 μm, P_{max} in the journal part increases by 22.6%, and in the thrust part, it increases by 16.1%, demonstrating the most significant influence.

Figure 3.24a illustrates T_{max} of the two models under four different radial clearances. As the radial clearance decreases, the film temperature in each part increases. The trend of T_{max} change mirrors that of P_{max}, with the main distinction being that T_{max} of each part increases when the radial clearance decreases in the proposed model, and the coupled effect is most pronounced at low radial clearances. This is attributed to the reduction in radial clearance, leading to a decrease in end discharge flow and an increase in frictional heat generation. Figure 3.24b presents the specific distributions of the proposed film temperatures in each part at time T/6, and the proposed T_{max} values are displayed in Fig. 3.25a. As c increases from 100 μm to 400 μm, T_{max} in each part decreases. The change rates indicate that, when using the proposed model, the influence of radial clearance on temperature is more pronounced under small clearances.

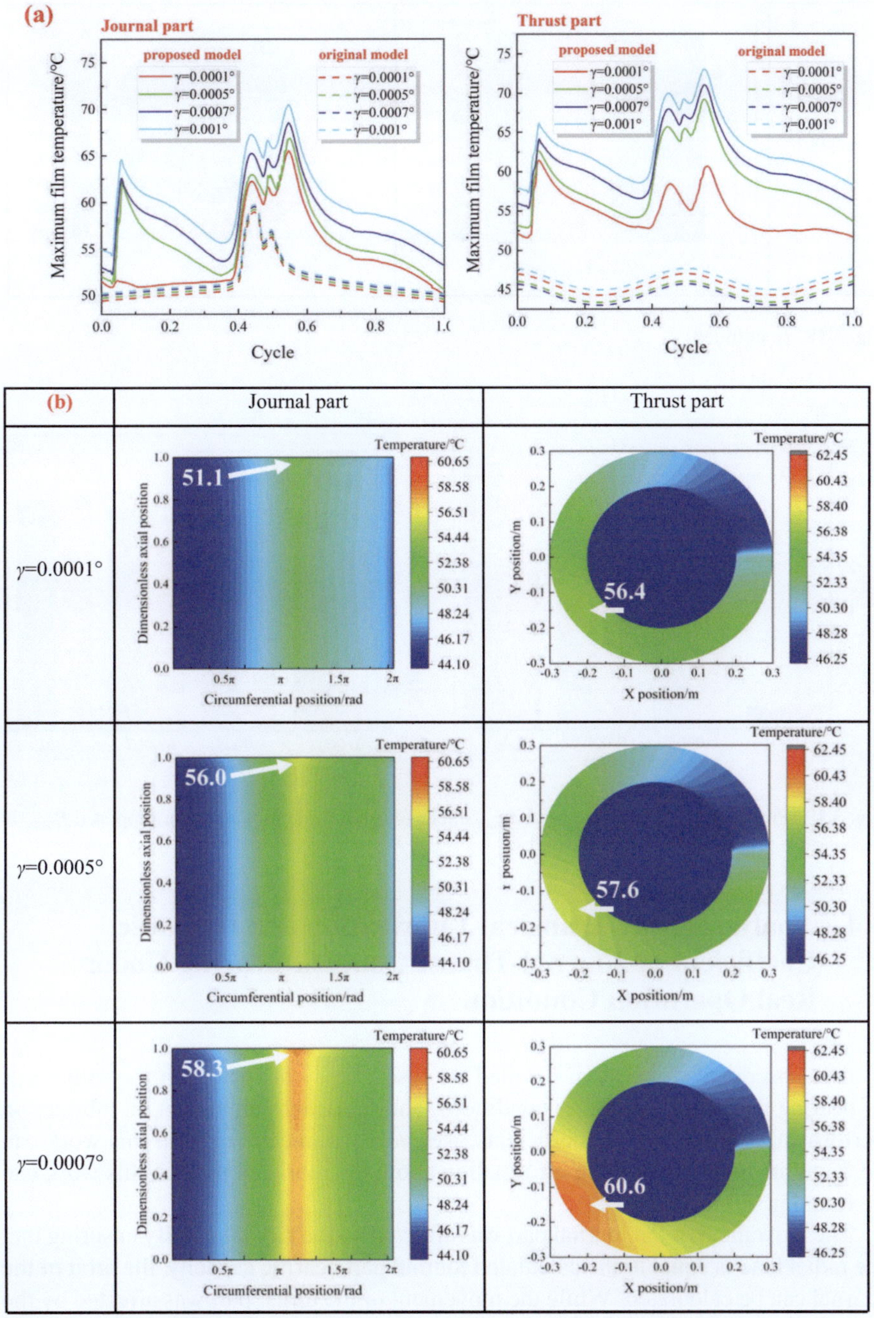

Fig. 3.18 T_{max} of each part under different incline angles: **a** proposed T_{max} and original T_{max} within a working cycle; **b** proposed film temperature distribution at T/6

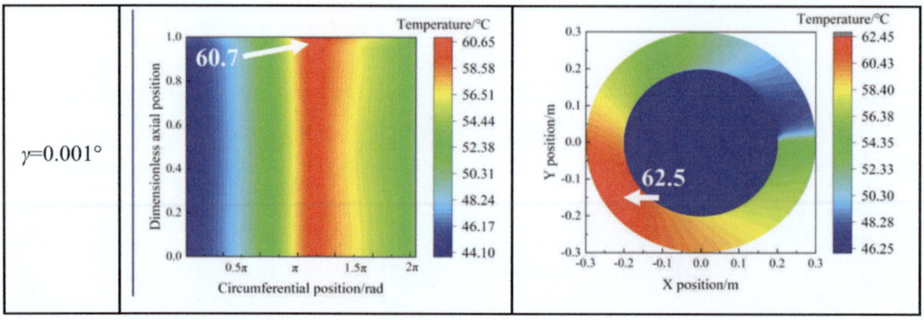

Fig. 3.18 (continued)

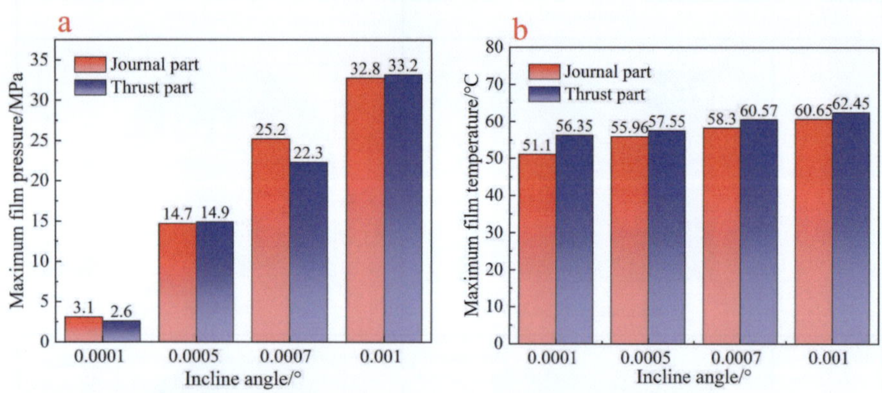

Fig. 3.19 Proposed P_{max} and proposed T_{max} under four incline angles at T/6: **a** proposed P_{max} in each part; **b** proposed T_{max} in each part

3.4 Analysis of the Transient Lubrication and Dynamic Coefficient of Journal-Thrust Coupled Bearing Under Real Operation Condition

In the subsequent discussion, detailed analyses were conducted on the influences of bearing geometry, operating conditions, and lubricant parameters on lubrication performance and dynamic coefficients. Relevant parameters used in this work can be found in Table 3.5 and the information about the lubricant used in this work can be found in Table 3.6.

The movement of the journal part was affected by the radial load. By ensuring that the radial load is equal to the calculated journal part bearing capacity, the orbit of the journal can be calculated. While the movement of the thrust part was affected by the axial shifting displacement, so the film thickness could be directly calculated, and then the film pressure and other lubrication parameters can be solved. Figure 3.26

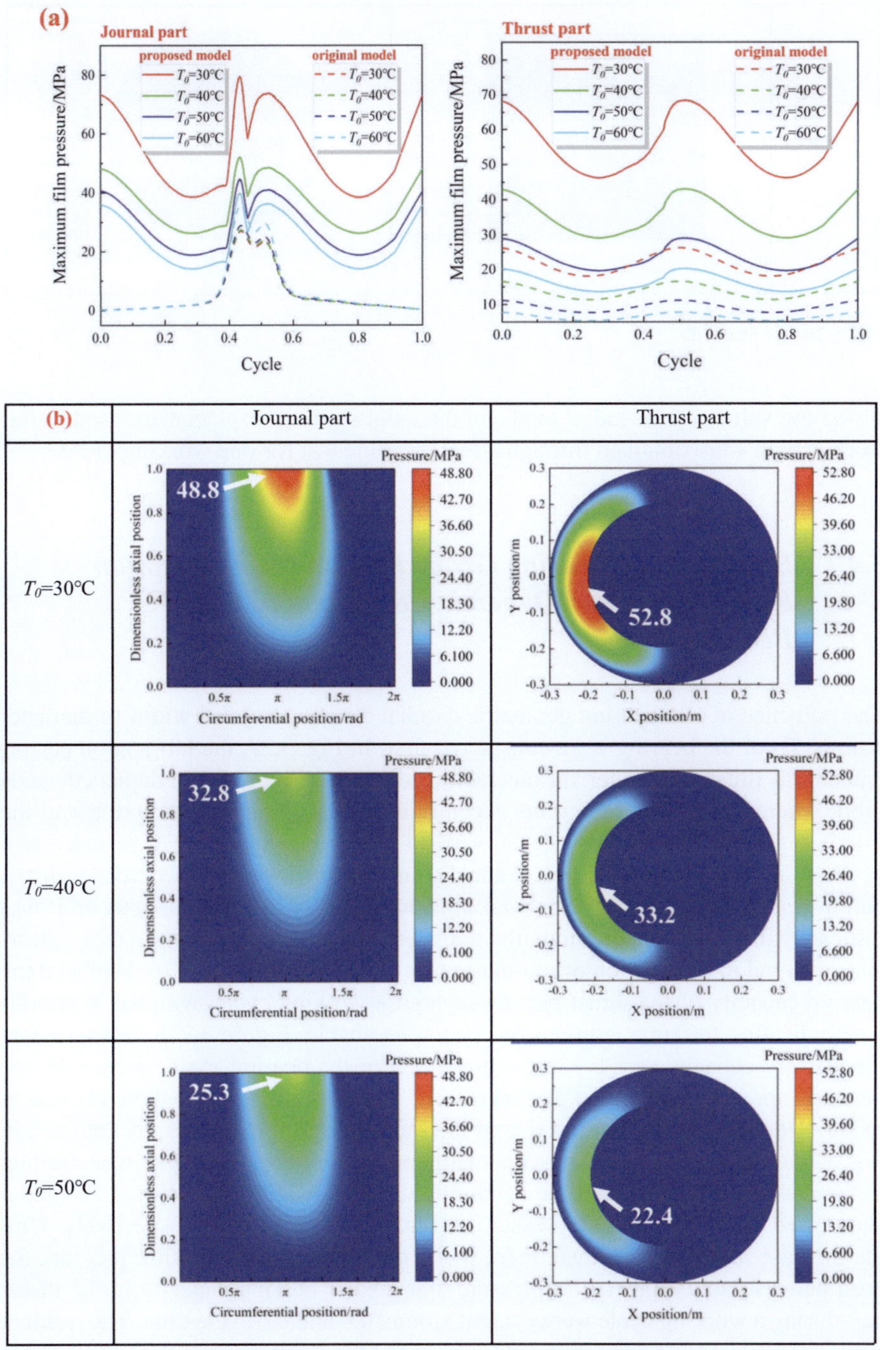

Fig. 3.20 $P_{\max}$ of each part under different inlet oil temperatures: **a** proposed $P_{\max}$ and original $P_{\max}$ within a working cycle; **b** proposed film pressure distribution at T/6

T_0=60°C

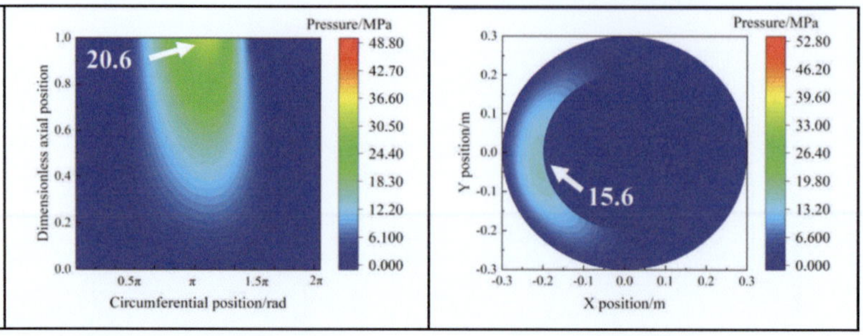

Fig. 3.20 (continued)

shows the values of the radial load and the axial shifting displacement used in this work, which were obtained through a real machine test for one working cycle.

3.4.1 Influences of Bearing Geometries on the Lubrication Performance and Dynamic Coefficient of the JTC Bearing

The influence of two bearing geometries: radial clearance(c) and width-to-diameter ratio (L/D) of the bearing were considered first. In Fig. 3.27, the lubrication performance was illustrated under various radial clearances. Figure 3.27a depicts the axis orbit, where the X-axis represents eccentricity in the horizontal direction and the Y-axis represents eccentricity in the vertical direction. Figure 3.27a revealed that as the C increased, the axis orbit gradually shifted downward, and a broader distribution was observed in the horizontal direction. The variation in the axis orbit was intricately linked to the minimum film thickness, meaning that the rise in radial clearance resulted in a reduction of minimum film thickness. Figure 3.27b displayed the bearing capacity of the thrust part throughout a working cycle, with the X coordinate indicating the time required for a complete cycle. Figure 3.27b indicated that the primary capacity area lay in the lower part of the bearing. As C decreased, the bearing capacity of the thrust part increased. This was attributed to the reduction in radial clearance, leading to the rise in the minimum film thickness, as depicted in Fig. 3.27a. Consequently, the maximum film pressure of the journal part rose, leading to the rise of film pressure at the common interface. Additionally, Fig. 3.27c and d showcased the maximum film pressure of journal and thrust parts, respectively. With the increase in radial clearance, there was a rise in the maximum film pressure for each part. Notably, higher film pressure and greater bearing capacity in the thrust part during a working cycle were evident around 0.4 and 0.6 cycle time. The result at around 0.4 cycle time corresponded to the radial load curve, while the results at the 0.6 cycle time were influenced by the coupled effect, where the load curve and axial movement displacement collectively contributed to the increase in value. Similarly, slight fluctuations could be observed at 0.1 and 0.7 cycle time.

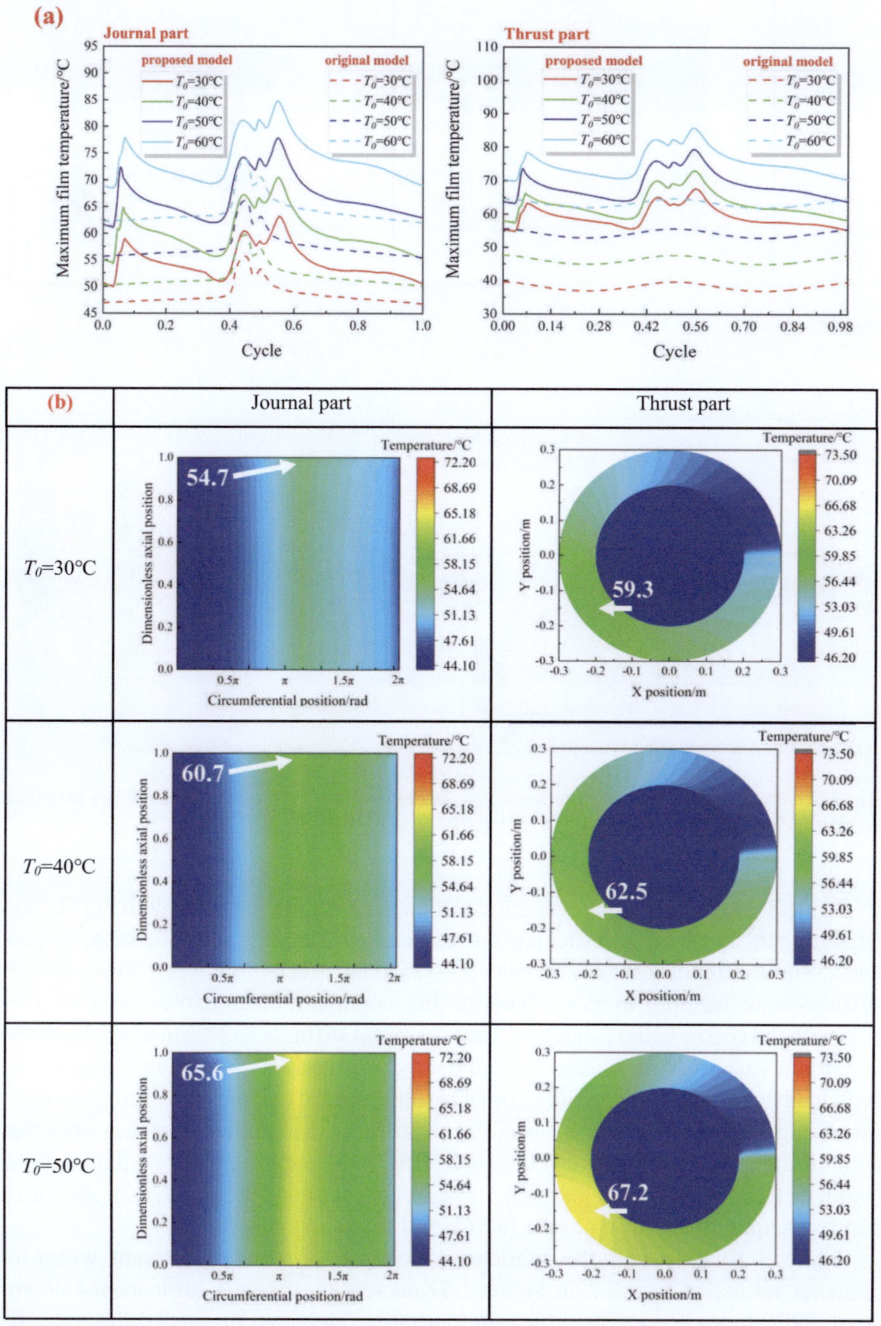

Fig. 3.21 $T_{\max}$ of each part under different inlet oil temperatures: **a** $T_{\max}$ within a working cycle; **b** proposed film temperature distribution at T/6

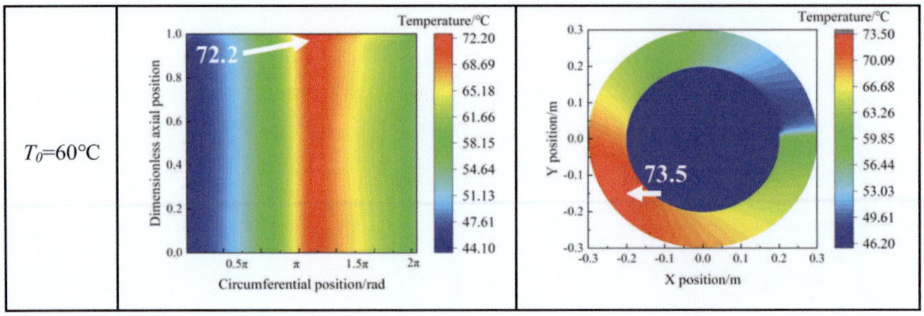

Fig. 3.21 (continued)

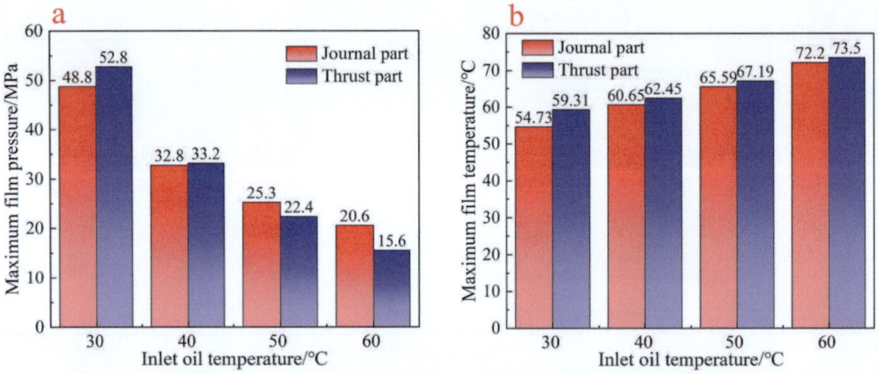

Fig. 3.22 Proposed $P_{\max}$ and the proposed $T_{\max}$ under four inlet oil temperatures at T/6: **a** proposed $P_{\max}$ in each part; **b** proposed $T_{\max}$ in each part

Figure 3.28 illustrates the dynamic coefficient of the JTC bearing, with Fig. 3.28a–f detailing the horizontal, vertical, axial stiffness and damping of the bearing, respectively. Similar trends were observed in both stiffness and damping behaviors. Taking stiffness as an example, there was a noticeable increase of the maximum bearing stiffness with the rise in radial clearance. For horizontal stiffness and vertical stiffness, the maximum stiffness was evident around 0.4 cycle time. This can be attributed to the peak load occurring at this point in time, resulting in the calculated disturbance pressure being at its highest. In the case of axial stiffness, a slight increase was observed with the augmentation of radial clearance. This phenomenon can be attributed to the coupled effect, wherein the journal part begins to influence the pressure of the thrust part, consequently resulting in the increase of the thrust part capacity.

Figure 3.29 illustrates the lubrication performance under different width-to-diameter ratios. In Fig. 3.29a, as the *L*/*D* decreases, the axis orbit moves downward, indicating a reduction in the minimum film thickness. Figure 3.29b shows the increase of the bearing capacity of the thrust part with a reduction in the *L*/*D*. Moreover, the difference between the bearing capacities at approximately 0.4 cycle time

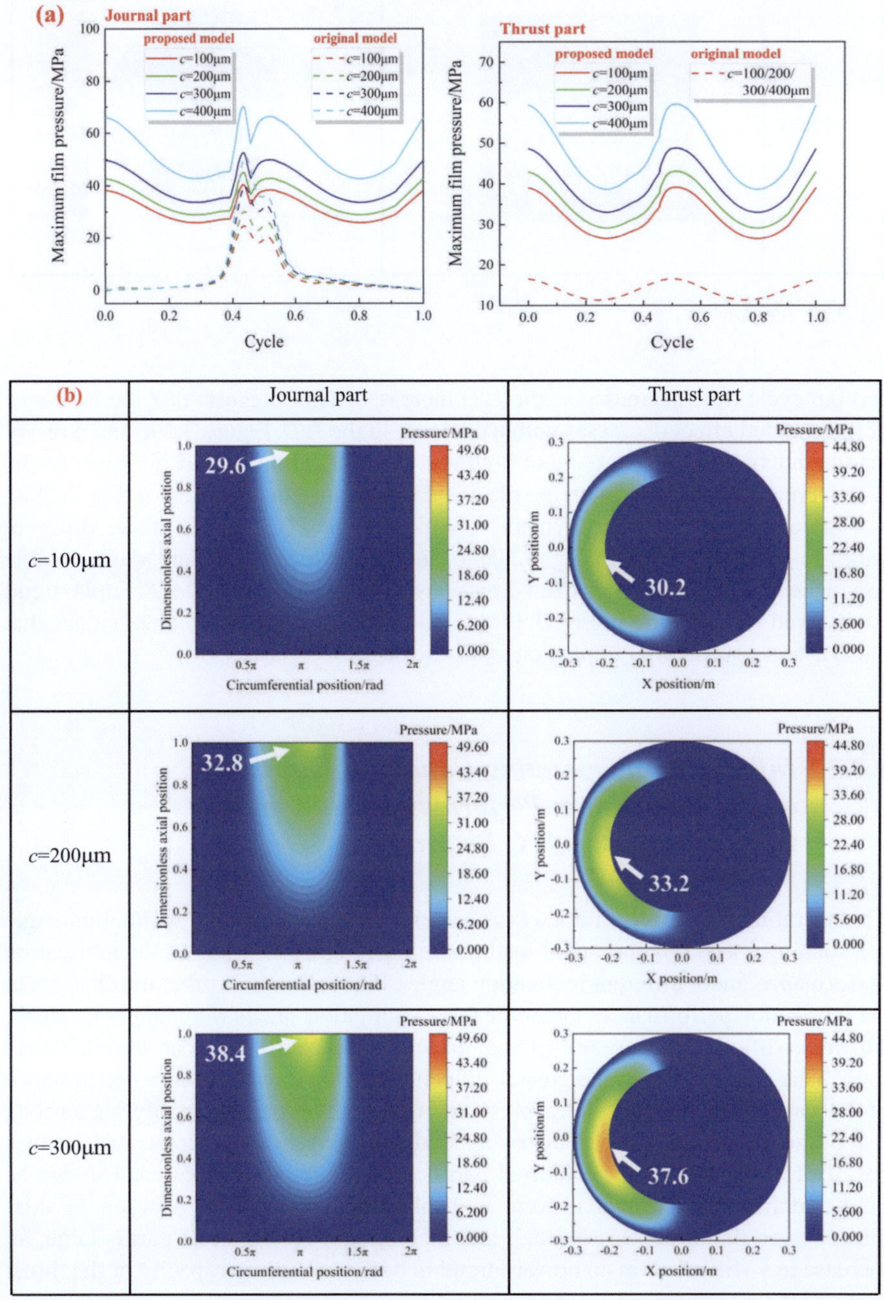

Fig. 3.23 $P_{\max}$ of each part under different radial clearances: **a** the proposed $P_{\max}$ and original $P_{\max}$ within a working cycle; **b** proposed film pressure distribution at T/6

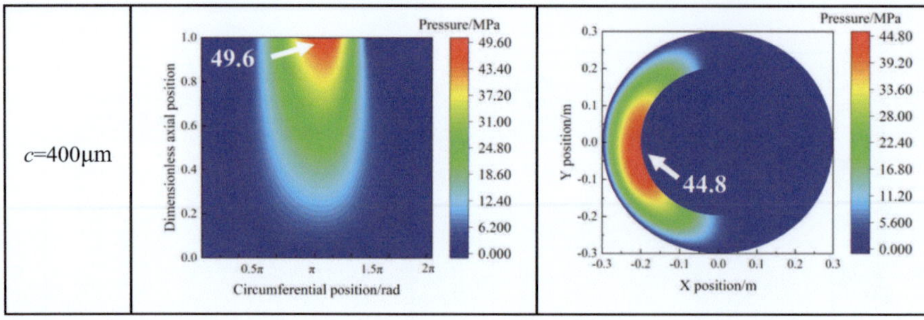

Fig. 3.23 (continued)

and 0.6 cycle time decreases as the *L/D* increases. This suggests that the influence of the coupled effect decreases with a decrease in the *L/D*. Figure 3.29c and d reveal that the maximum film pressure of both journal part and thrust part increases as the *L/D* decreases, consistent with the observed minimum film thickness in Fig. 3.29a.

Figure 3.30 presents the dynamic coefficients of JTC bearing under different width-to-diameter ratios. In Fig. 3.30a–c, as the *L/D* increases, there was a notable rise in the maximum stiffness in all directions within a working cycle. A similar trend is observed for damping in Fig. 3.30d–f. Additionally, Fig. 3.30c demonstrates that the *L/D* has a substantial impact on axial stiffness.

3.4.2 Influences of Operating Condition Parameters on the Lubrication Performance and Dynamic Coefficient of the JTC Bearing

The operating condition parameters considered in this study were inclination angle (γ), rotating speed (ω) and radial load ($Load_{max}$). Figure 3.31 shows the lubrication performance under different inclination angles. The results show that the changes in the lubrication performances under the three inclination angles were relatively small. The main differences were reflected at around 0.1 cycle time and around 0.7 cycle time. This is attributed to the greater influence of γ on the lubrication performance of the thrust part. The calculated pressure of the journal part is notably higher than that of the thrust part by using the radial load and axial shifting displacement values specified in this study, so the journal part dominate the lubrication result. It can be predicted that when the axial movement displacement is large, that is, when the axial clearance on the thrust part is small and the pressure on the thrust part is large, an increase in γ will result in an upward trend in both the bearing capacity of the thrust part and the maximum pressure of each part.

Figure 3.32 illustrates the dynamic coefficient of the JTC bearing under varying inclination angles. Similar to the trend of lubrication performance, the maximum horizontal stiffness and vertical stiffness within a working cycle under different

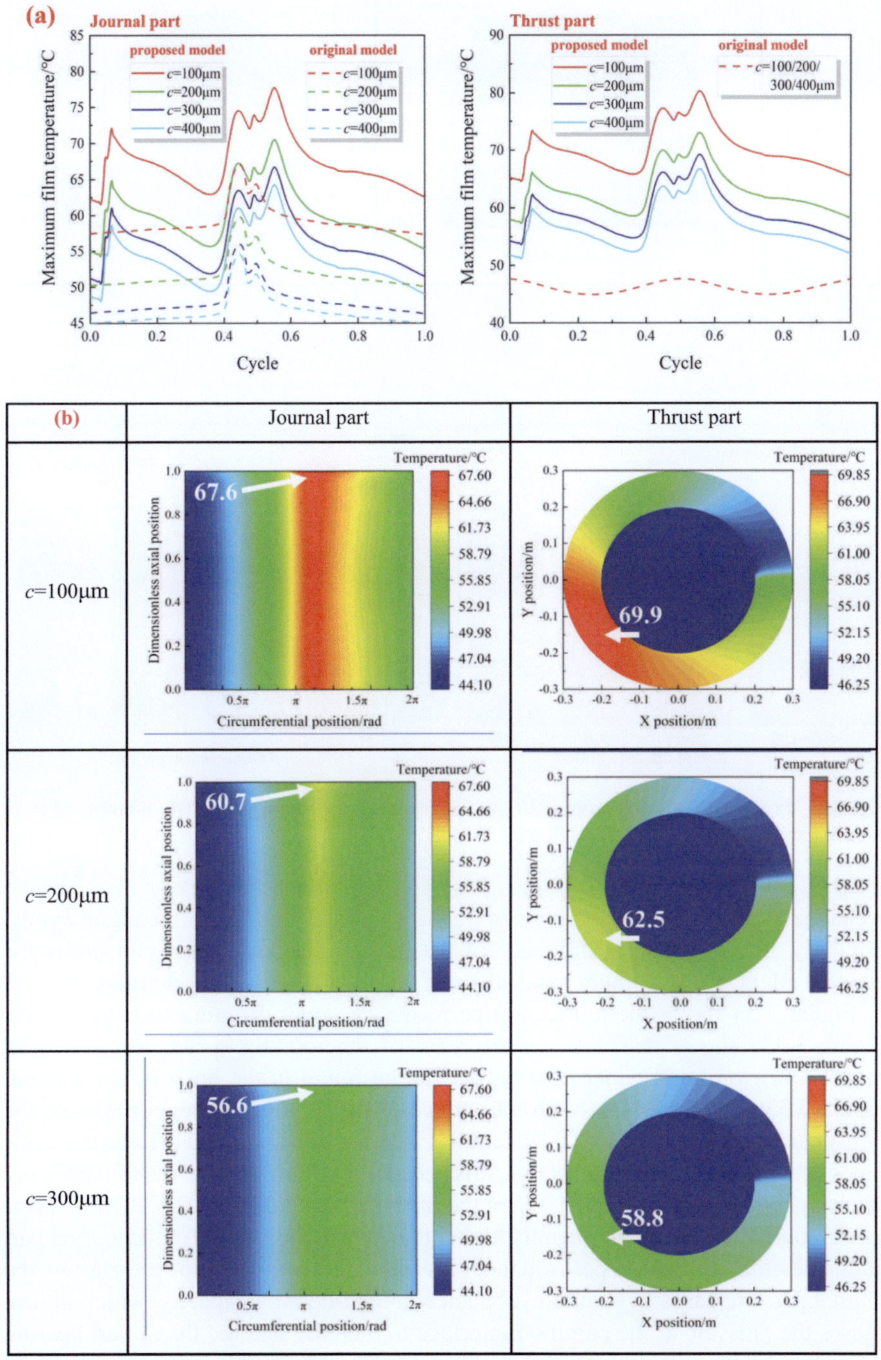

Fig. 3.24 P_{max} of each part under different radial clearances: **a** proposed T_{max} and original T_{max} within a working cycle; **b** proposed film temperature distribution at T/6

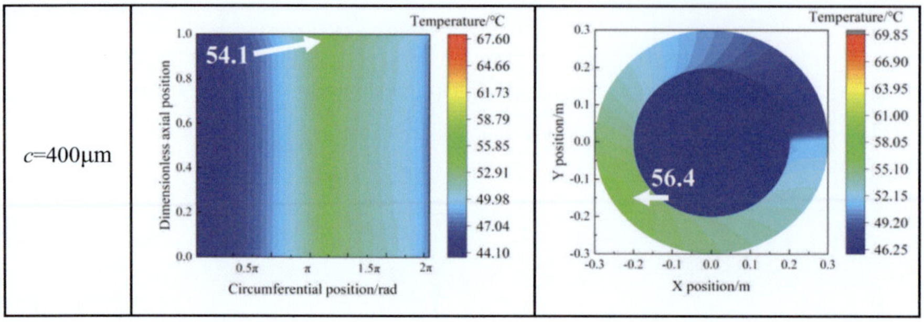

Fig. 3.24 (continued)

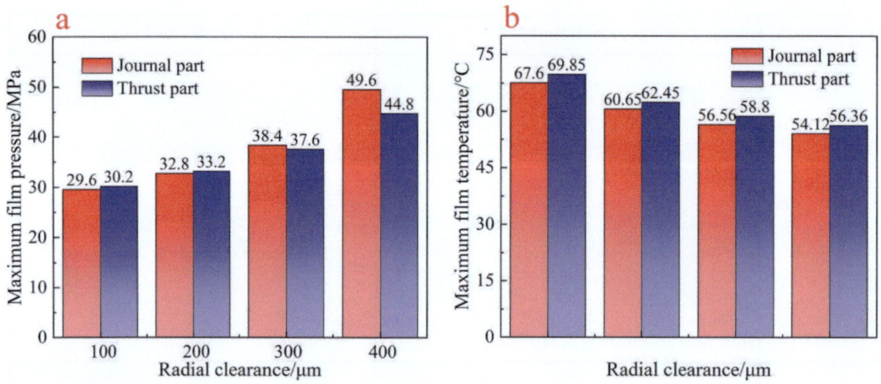

Fig. 3.25 Proposed P_{max} and proposed T_{max} under four radial clearances at T/6: **a** proposed P_{max} in each part; **b** proposed T_{max} in each part

γ change slightly. As the increase of γ, the axial stiffness increases significantly. Similarly, due to the joint influence of the journal part and thrust part under the coupled effect, the maximum axial stiffness occurs at about 0.5 cycle time.

Figure 3.33 shows the lubrication performance under different rotating speeds. Figure 3.33a shows that with the decrease of the rotating speed, the axis orbit moves downward and wider distribution can be found in the horizontal direction. Figure 3.33b illustrates that with the reduce of the ω, the bearing capacity of the thrust part at around 0.4 and 0.6 cycle times increase and at around 0.1 and 0.7 cycle times decrease. As mentioned above, this because at different time, the film pressure at each part was quite different. When the time was around 0.4 or 0.6, the film pressure at journal part was significantly higher than that at thrust part, so the journal part dominates the lubrication performance. As the ω reduces, the film pressure of the journal part increases to maintain a balance in load-bearing capacity, which in turn causes the pressure at the common interface to increase and the thrust part bearing capacity increase. When the time was around 0.1 or 0.7, the film pressure at journal part was significantly lower than that at thrust part, so the thrust part dominates the

Table 3.5 Parameters used in this work

Parameter	Value
Radius clearance/c	150/200/300 μm
Axial clearance/h_0	250 μm
Journal width/L	0.19/0.16/0.12 m
Inclination angle of journal/γ	0.01/0.008/0.012°
Supply temperature/T_0	30/40/50 °C
Elastic modulus/E_J	105GPa
Poisson ratio/v_p	0.327
Rotating speed of the crankshaft/ω	400/500/600 rpm
Viscosity temperature coefficient/β	– 0.0491
Thermal expansion coefficient of lubricant/α	7.14E–4/K
Density of the bearing/ρ_b	8950 kg/m^3
Heat transfer coefficient/h_b	80 W/m^{2}K
Heat capacity of lubricant/c_{pl}	2,000 J/°C
Heat capacity of bearing/c_{pb}	515 J/°C
Thermal conductivity of bearing /k_b	47 W/mK
Thermal conductivity of lubricant/k_l	0.13 W/mK
Thermal conductivity of journal/k_J	51.9W/mK

Table 3.6 Lubrication parameters

Parameter	Value
Lubricant grade	CD15W-40
Kinematic viscosity (100°C)	15.03 mm^2/s
Viscosity Index	134
Flash point temperature	232 °C
Pour point temperature	– 32 °C
Base value	9.06mgKOH/g
Sulfated ash (mass fraction)	1.1%
chemical composition	Mineral oil, Calcium sulfonate, Zinc dialkyldithiophosphate (ZDDP), Succinimide compounds, Benzotriazole, Polymethacrylates

lubrication performance. With the increase of the ω, the film pressure of thrust part increase according to the Reynolds equation. The trends of film pressure in journal and thrust parts were also similar as shown in Figure 3.33c and d.

Figure 3.34 shows the dynamic coefficient of JTC bearing under different rotating speeds. With the increase of the ω, the maximum stiffness at each direction within a working cycle increase. The stiffness change was relatively linear, and the maximum radial stiffness appears at around 0.4 cycle time, which corresponds to the position

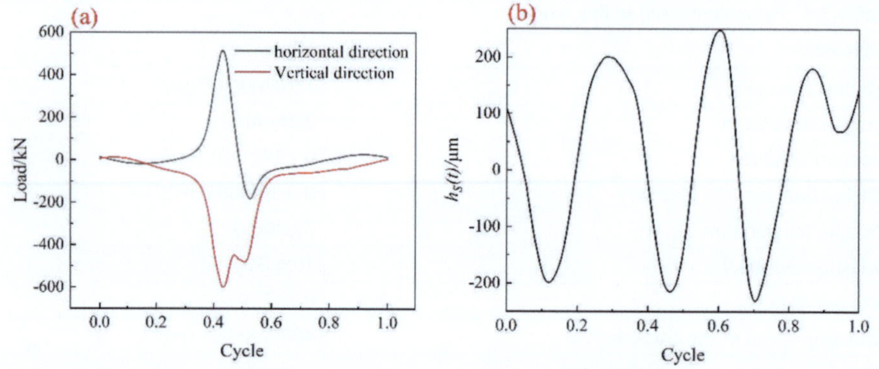

Fig. 3.26 The input parameters: **a** radial load; **b** axial shifting displacement of the crankshaft

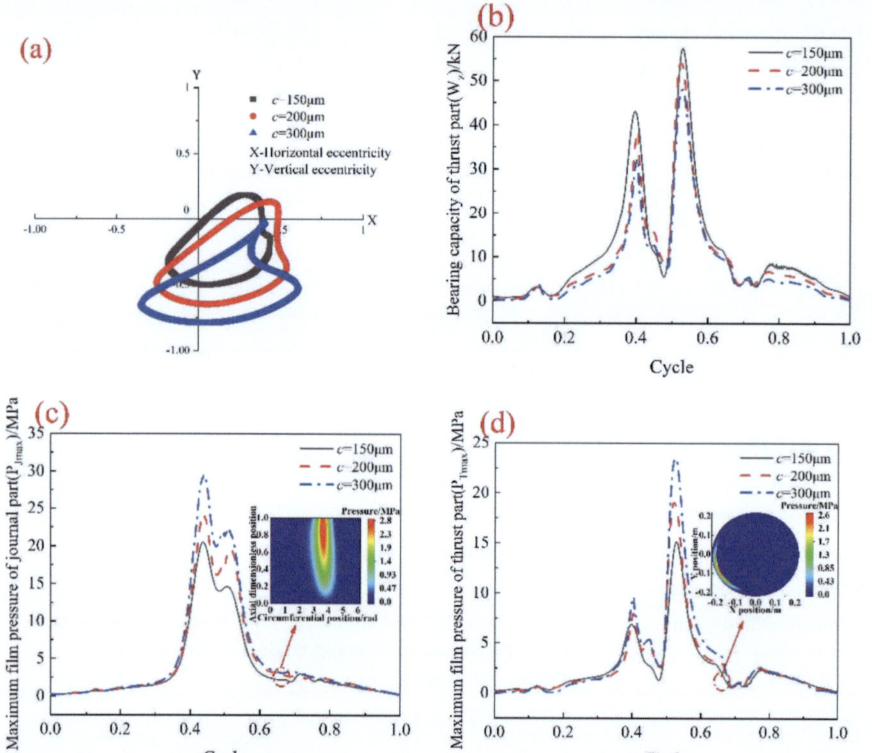

Fig. 3.27 Influence of the radial clearance on the lubrication performance of the JTC bearing: **a** axis orbit of journal; **b** capacity of the thrust part; **c** maximum film pressure of the journal part; **d** maximum film pressure of the thrust part

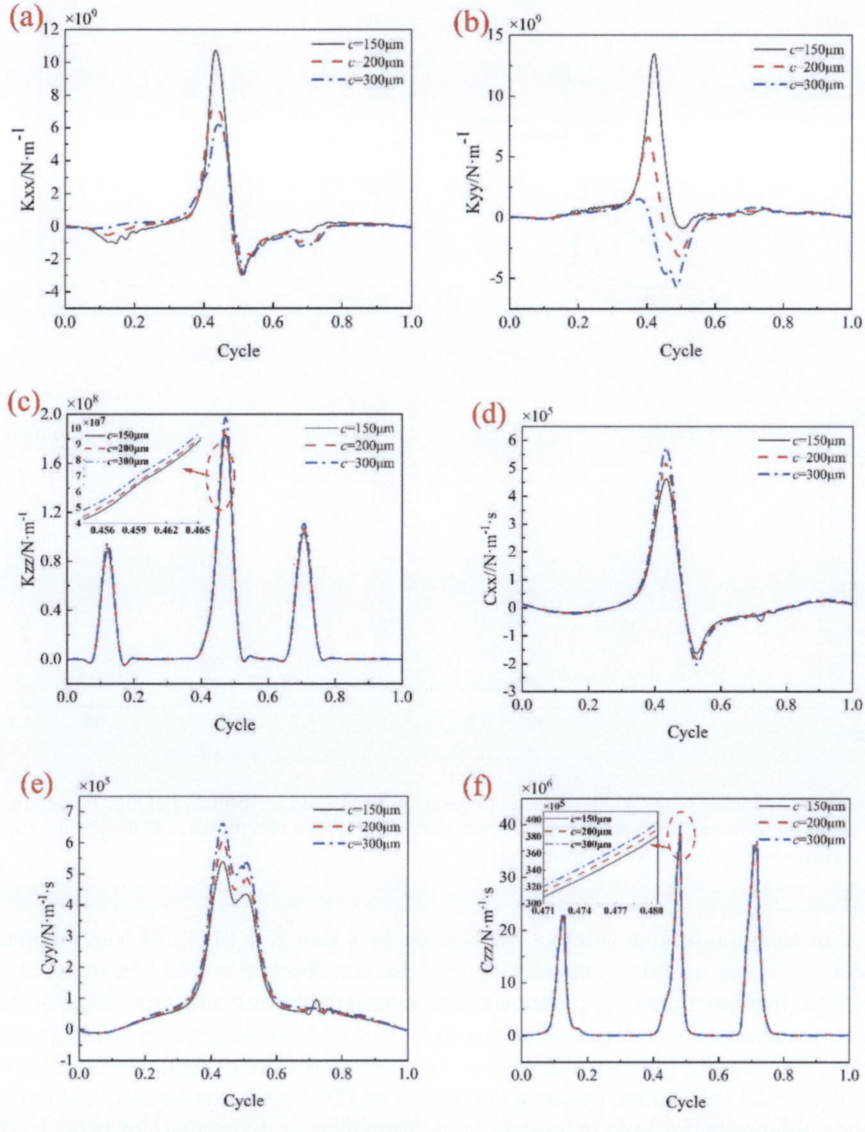

Fig. 3.28 Influence of the radial clearance on the dynamic coefficient of the JTC bearing: **a** *Kxx*; **b** *Kyy*; **c** *Kzz*; **d** *Cxx*; **e** *Cyy*; **f** *Czz*

of the maximum load; the maximum axial stiffness appears at around 0.5, since the radial load was relatively large and the axial clearance of the thrust part was minimal at this time.

Figure 3.36 shows the lubrication performance under different radial loads, and the loads used in this work were different in maximum value but the same in shape.

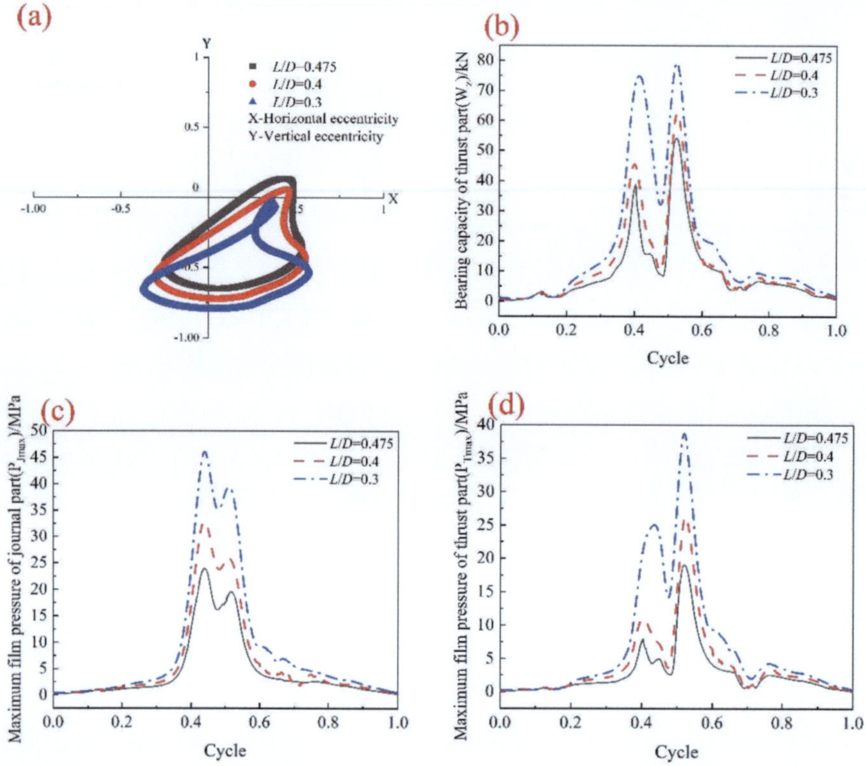

Fig. 3.29 Influence of the width-diameter ratio on the lubrication performance of the JTC bearing: **a** axis orbit of journal; **b** capacity of the thrust part; **c** maximum film pressure of the journal part; **d** maximum film pressure of the thrust part

The distribution of load within a working cycle is shown in Fig. 3.35, and the load here is a scalar quantity, representing the resultant force. Figure 3.36a shows that with the increase of $Load_{max}$, the axis orbit moved downward while the shape of the orbit remain the same. Figure 3.36b–d show that with the increase of $Load_{max}$, the thrust part bearing capacity and the maximum film pressure of each part increased.

Figure 3.37 shows the dynamic coefficient of JTC bearing under different radial loads. Similar to the trend in lubrication performance, as the increase of radial load, the maximum stiffness in all directions increased. It can be found that although the radial load is the input parameter in journal part, considering coupled effect making it influenced the axial stiffness as well.

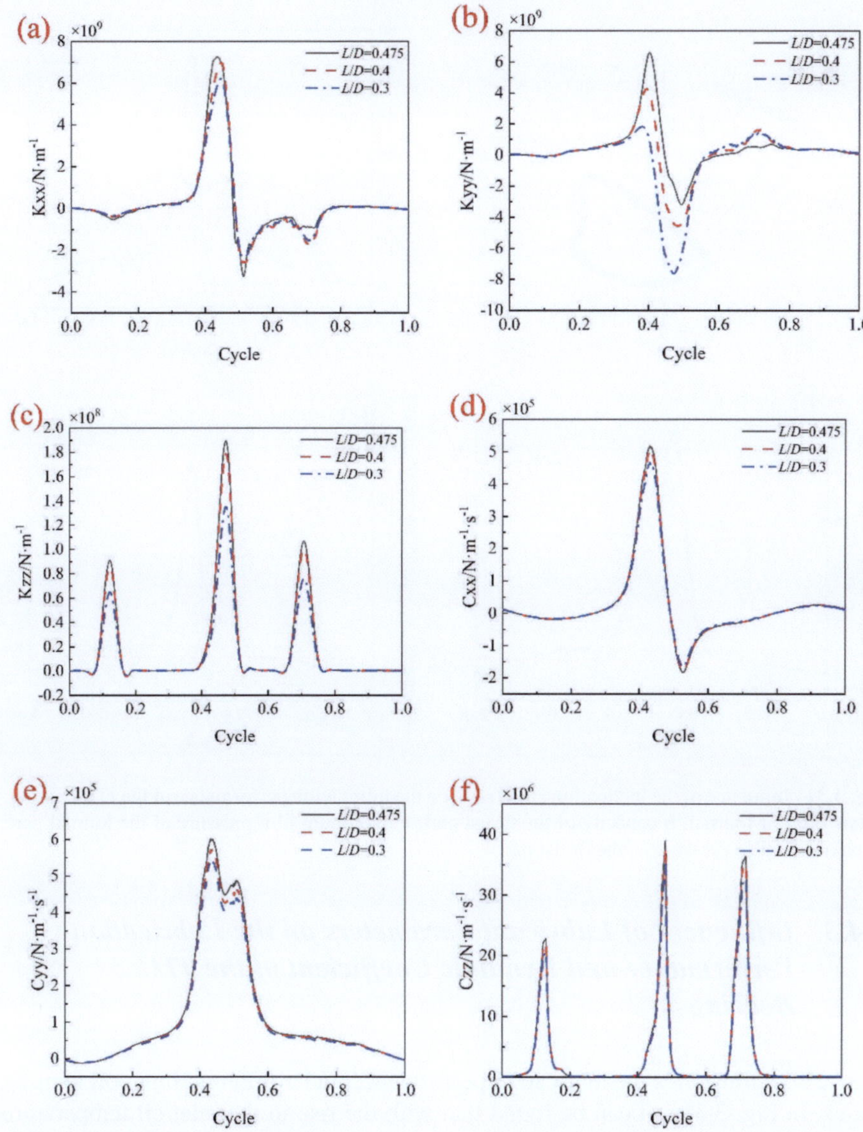

Fig. 3.30 Influence of the width-diameter ratio on the dynamic coefficient of the JTC bearing: **a** *Kxx*; **b** *Kyy*; **c** *Kzz*; **d** *Cxx*; **e** *Cyy*; **f** *Czz*

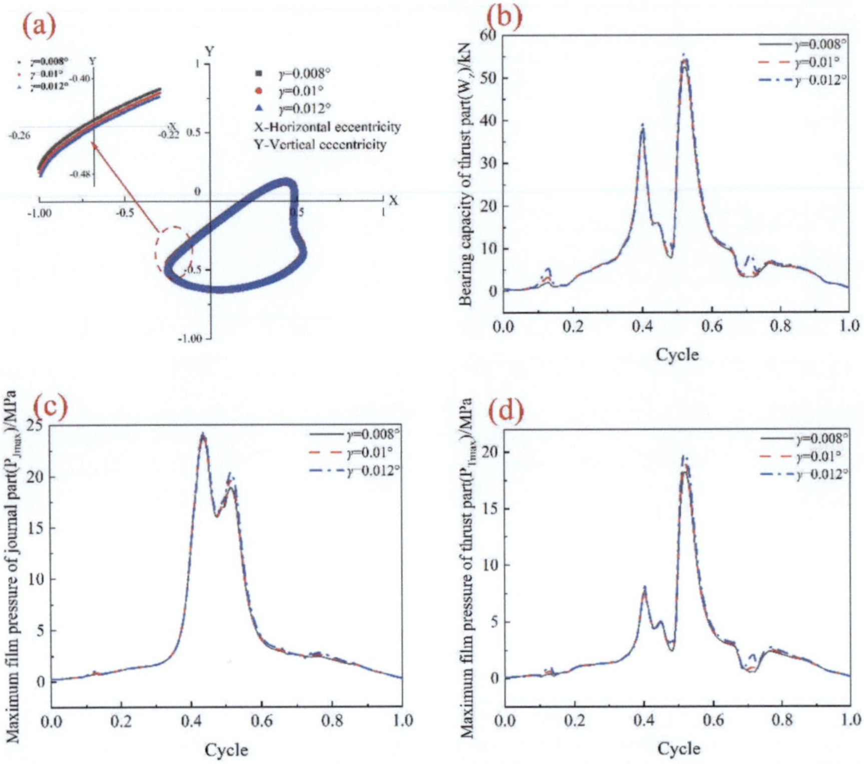

Fig. 3.31 Influence of the inclination angle ratio on the lubrication performance of the JTC bearing: **a** axis orbit of journal; **b** capacity of the thrust part; **c** maximum film pressure of the journal part; **d** maximum film pressure of the thrust part

3.4.3 Influences of Lubricant Parameters on the Lubrication Performance and Dynamic Coefficient of the JTC Bearing

Figure 3.38 illustrates the lubrication performance under different input oil temperatures. In Fig. 3.38a, it can be found that with the rise in the inlet oil temperature (T_{in}), the axis orbit moves downward and has a wider distribution in the X direction. Additionally, Fig. 3.38a–d show that as the rise in T_{in}, the maximum bearing capacity of thrust part and the maximum film pressure of each part within a working cycle increased. The opposite trend occurred at around 0.1 and 0.7 cycle times because the influence of the thrust part in the lubrication performance was dominant at this time. With the increase in T_{in}, the viscosity of the oil decreased. For the journal part, to obtain sufficient bearing capacity to satisfy the load balance, the film thickness of journal part reduced, leading to an increase in film pressure. Meanwhile, for thrust part, the film thickness was controlled by the axial shifting displacement, which was fixed. As the viscosity decreases, the pressure also decreased.

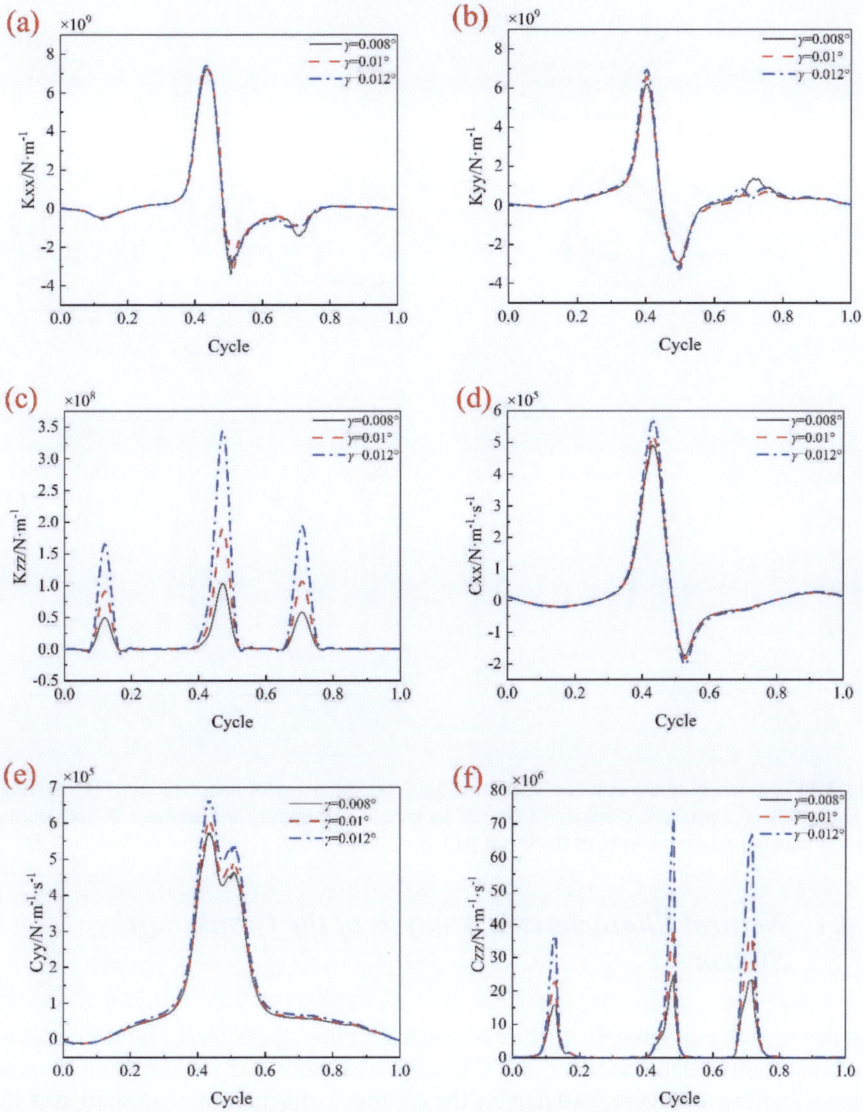

Fig. 3.32 Influence of the inclination angle on the dynamic coefficient of the JTC bearing: **a** *Kxx*; **b** *Kyy*; **c** *Kzz*; **d** *Cxx*; **e** *Cyy*; **f** *Czz*

Figure 3.39 illustrates the dynamic coefficient of JTC bearing under different input oil temperatures. Figure 3.39a–c show that with decrease of the T_{in}, and the stiffness in all direction increased. This is because the viscosity increases as the T_{in} rises, leading to the rise of the pressure. Also, the impact of temperature on axial stiffness was very significant with the consideration of coupled effect.

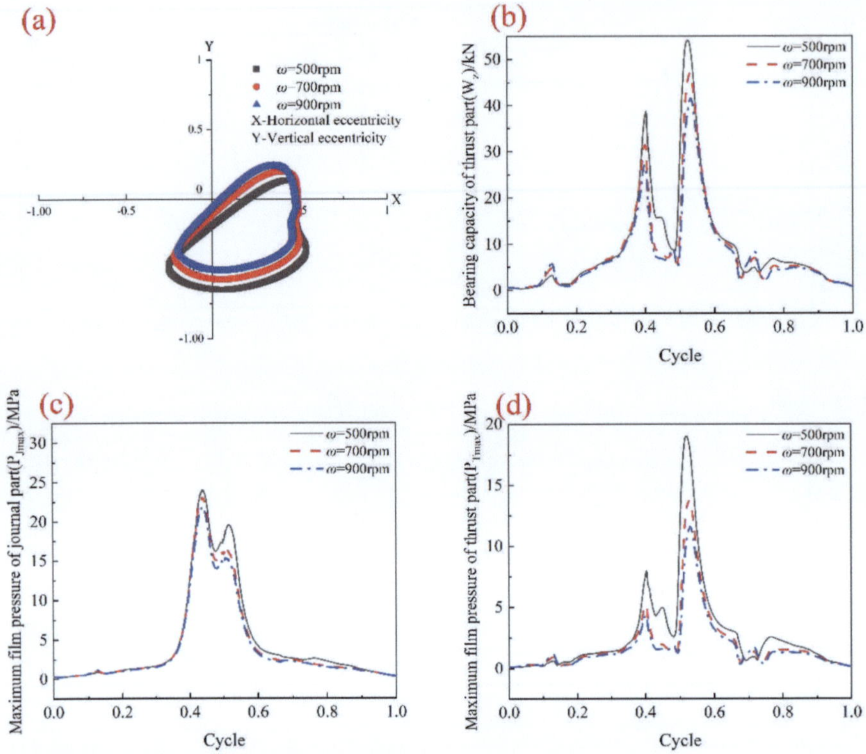

Fig. 3.33 Influence of the rotating speed on the lubrication performance of the JTC bearing: **a** axis orbit of journal; **b** capacity of the thrust part; **c** maximum film pressure of the journal part; **d** maximum film pressure of the thrust part

3.4.4 Natural Characteristic Analysis of the Crankshaft System

On the basis of obtaining the stiffness in each direction, the natural characteristics of the crankshaft system including a JTC bearing was analyzed based on finite element analysis software. Figure 3.40 depicts the schematic diagram of crankshaft system. Furthermore, the modal analysis of this system can be carried out, and the natural characteristics of the crankshaft system like natural frequency and vibration shapes of each order can be obtained. To emphasize the importance of considering the stiffness of JTC bearing in crankshaft systems, a necessity analysis was first performed. In the crankshaft model, the stiffness of the main bearing used was set to the value of 1E6 N/m, and the stiffness of the JTC bearing was selected from the results at 0.40694 cycle time which was calculated previously. Table 3.7 shows the natural frequencies of the first 10 orders in the model with and without considering stiffness values calculated for the JTC bearing, and Fig. 3.41 shows the comparison. The

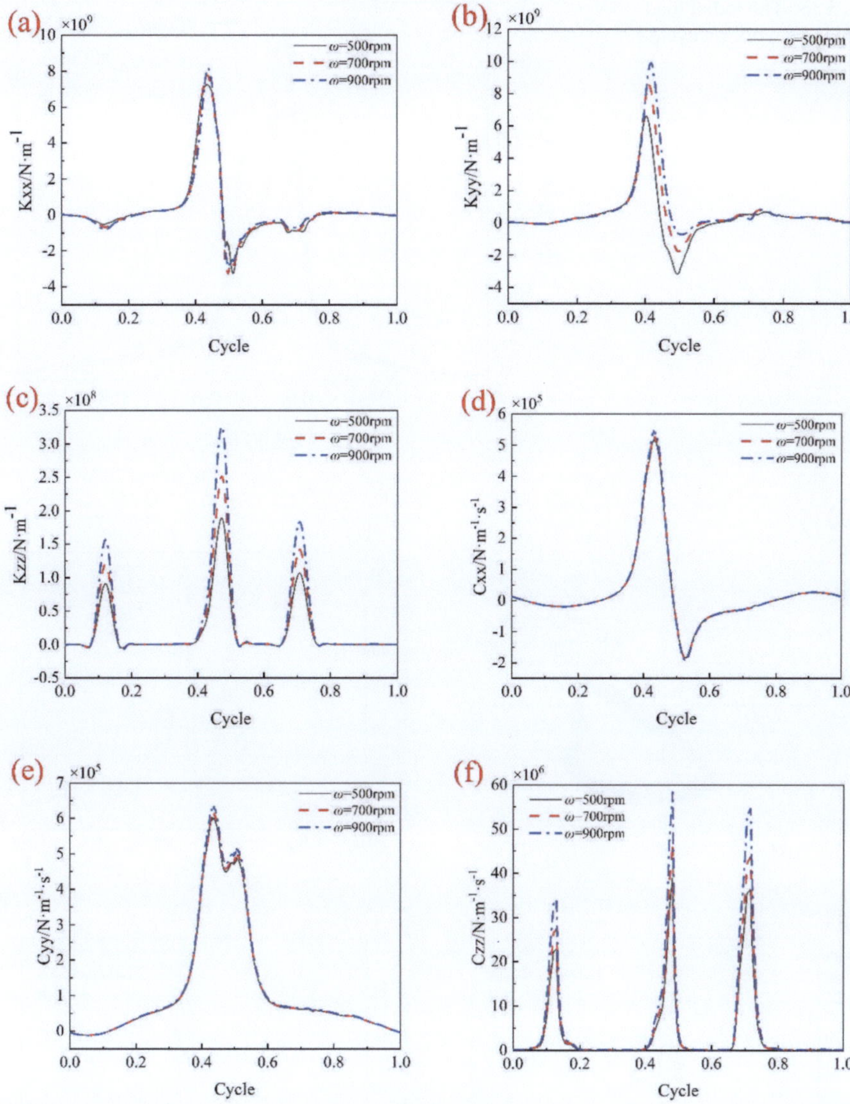

Fig. 3.34 Influence of the rotating speed on the dynamic coefficient of the JTC bearing: **a** *Kxx*; **b** *Kyy*; **c** *Kzz*; **d** *Cxx*; **e** *Cyy*; **f** *Czz*

results indicate significant differences in the natural frequencies calculated by the two models, with the largest deviation in the second order reaching 57.95%.

Figure 3.42 shows the first three-order vibration shapes of different models. Great differences can also be found in the vibration shape under various models. For the first-order vibration shape, as shown in Fig. 3.42a and d, when the JTC bearing was not considered in the model, the maximum displacement (MX) occurred at the last

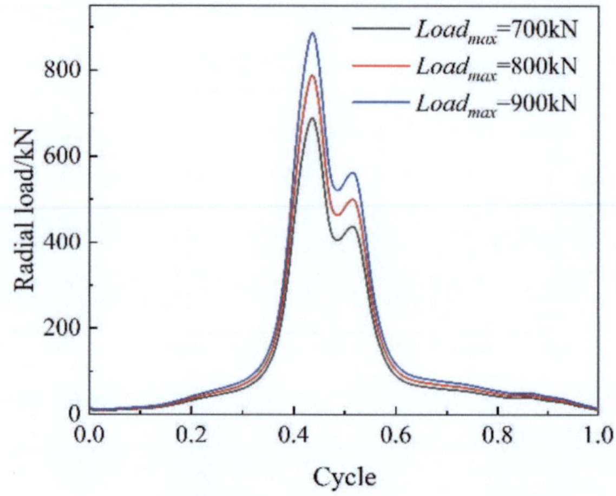

Fig. 3.35 The radial load within one working cycle

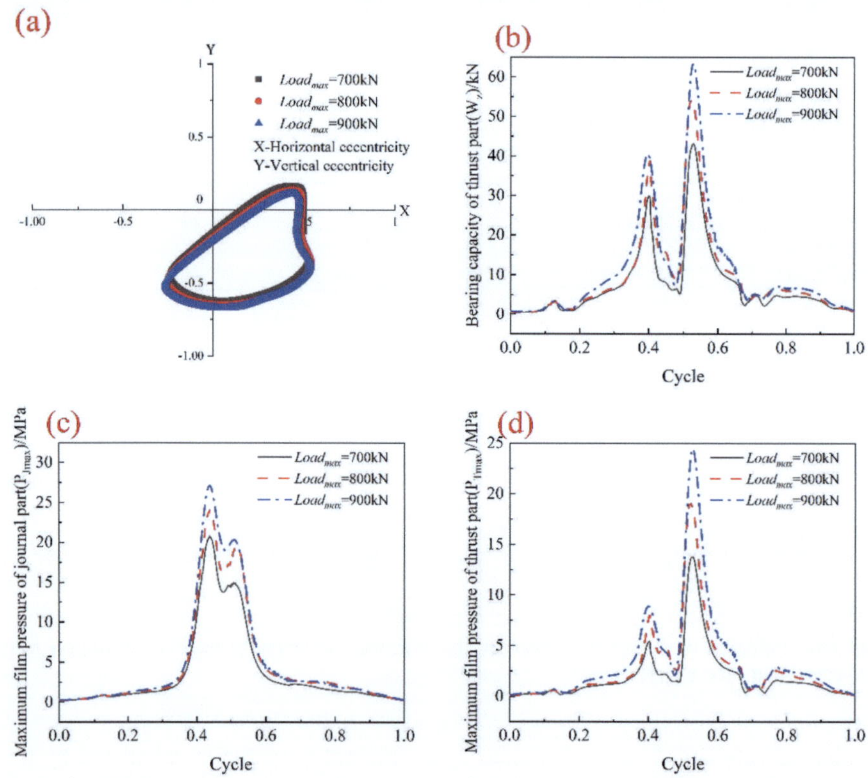

Fig. 3.36 Influence of the load on the lubrication performance of the JTC bearing: **a** axis orbit of journal; **b** capacity of the thrust part; **c** maximum film pressure of the journal part; **d** maximum film pressure of the thrust part

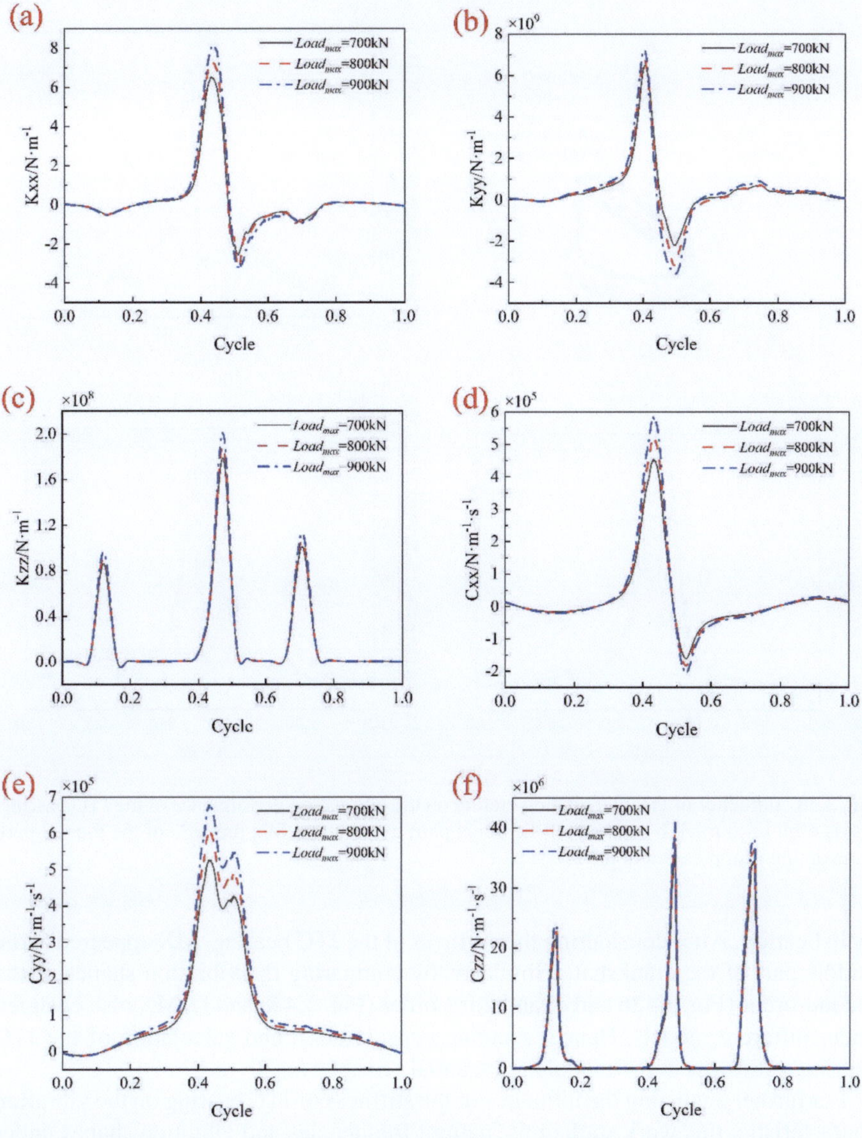

Fig. 3.37 Influence of the load on the dynamic coefficient of the JTC bearing: **a** *Kxx*; **b** *Kyy*; **c** *Kzz*; **d** *Cxx*; **e** *Cyy*; **f** *Czz*

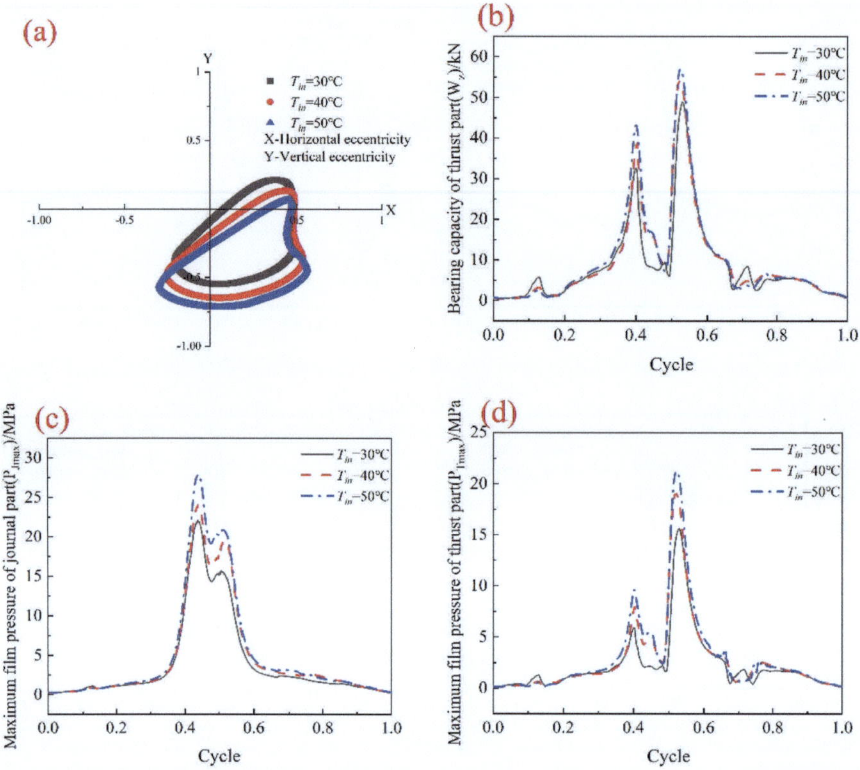

Fig. 3.38 Influence of the inlet oil temperature on the lubrication performance of the JTC bearing: **a** axis orbit of journal; **b** capacity of the thrust part; **c** maximum film pressure of the journal part; **d** maximum film pressure of the thrust part

main bearing. After considering the stiffness of the JTC bearing, MX appeared at the middle part of the crankshaft. Similarly, by comparing the vibration shapes of the second-order (Fig. 3.42b and e) and third-order (Fig. 3.42c and f), MX also changed under different models. Hence, accurate consideration and calculation of the JTC bearing's stiffness in the model are essential.

For further analyzing the influence of the stiffness of JTC bearing on the vibration characteristic, this work studied the natural frequencies and vibration shapes under different stiffness of the JTC bearing. To cover different extreme situations, three groups of stiffnesses with large differences were selected for calculation.

Figure 3.43a–c show the natural frequencies of the first 10 orders under the model with different horizontal stiffness, vertical stiffness and axial stiffness. The result shows that the changing trend of natural frequency with stiffness in each direction is similar, as the increase of the stiffness, the natural frequency rises, and is more obvious at high orders.

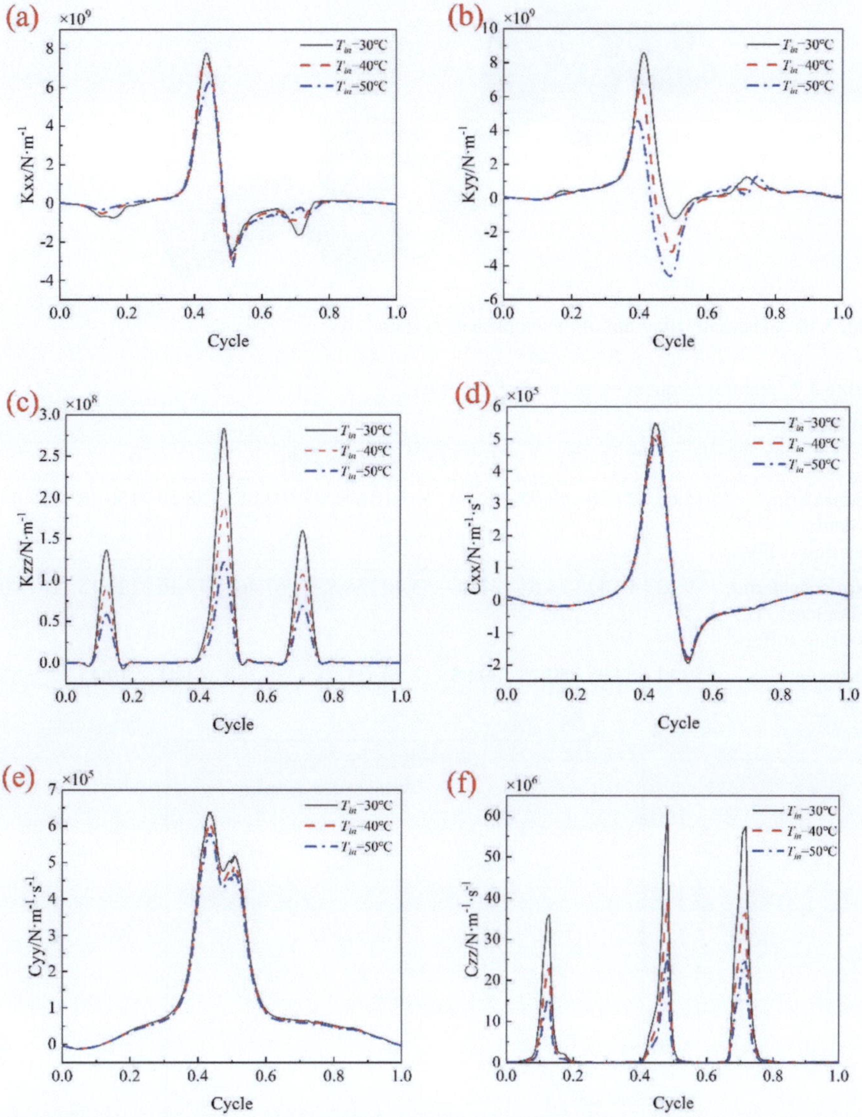

Fig. 3.39 Influence of the inlet oil temperature on the dynamic coefficient of the JTC bearing: **a** *Kxx*; **b** *Kyy*; **c** *Kzz*; **d** *Cxx*; **e** *Cyy*; **f** *Czz*

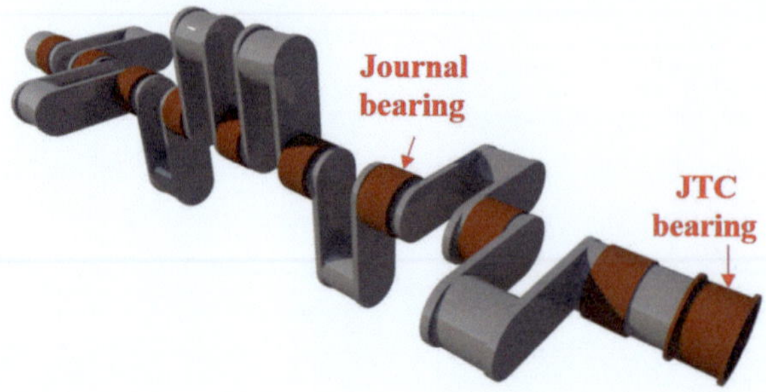

Fig. 3.40 Schematic diagram of the crankshaft system

Table 3.7 Natural frequencies of the first ten orders

Model	Order									
	1	2	3	4	5	6	7	8	9	10
Considering calculated JTC bearing stiffness	18.68	24.78	49.43	52.30	60.61	66.29	104.35	138.79	150.40	178.09
Not considering calculated JTC bearing stiffness	9.32	10.42	24.64	31.46	53.40	58.83	67.46	77.16	121.15	161.40
Difference/%	50.11	57.95	50.15	39.85	11.90	11.25	35.35	44.41	19.45	9.37

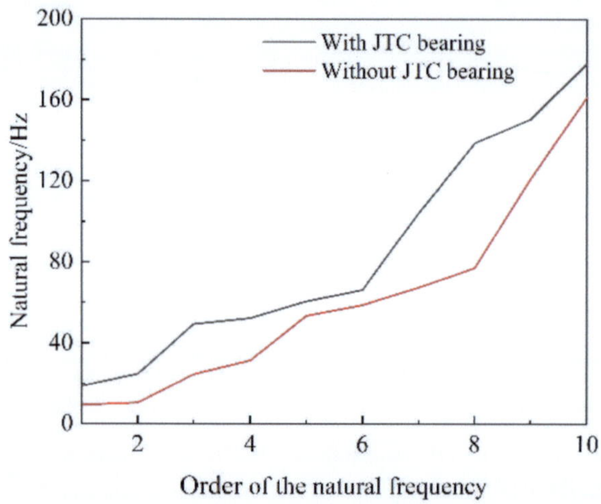

Fig. 3.41 Comparison of natural frequencies with and without consideration of JTC bearings in the system

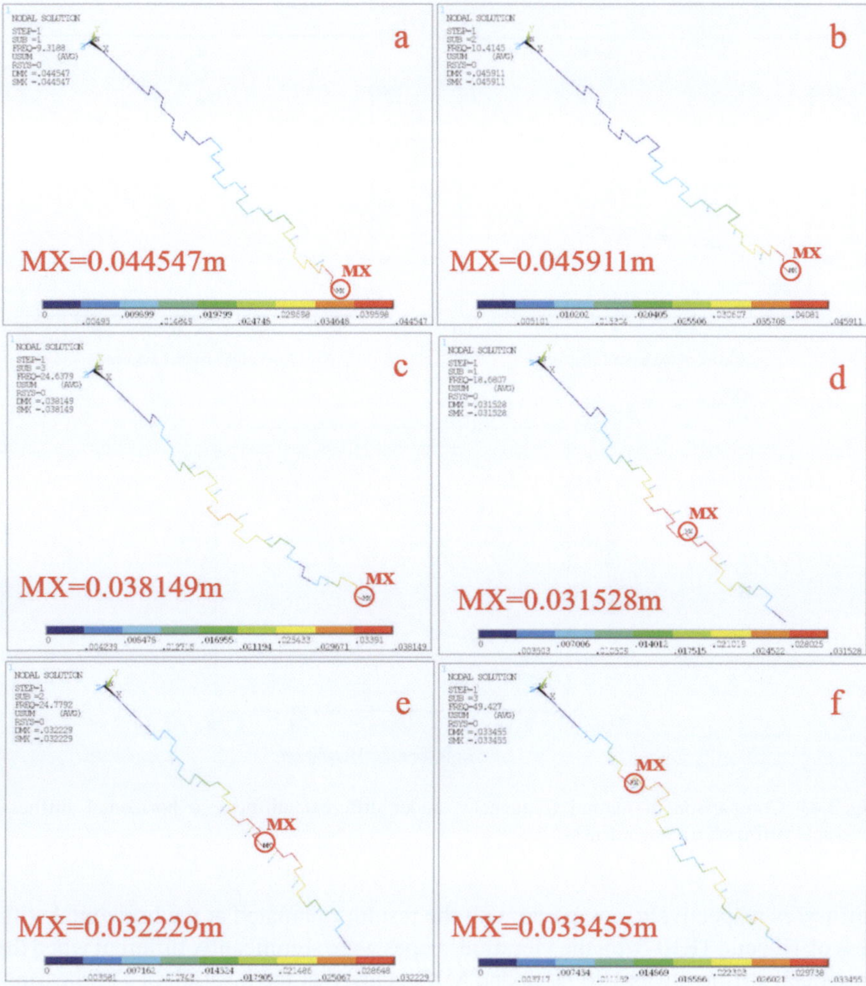

Fig. 3.42 Comparison of vibration shape under different model: **a–c** not considering JTC bearing; **d–f** considering JTC bearing

Furthermore, the vibration shapes under different stiffnesses were analyzed. Due to the crankshaft system studied in this work containing nine main bearings and one JTC bearing, when only the stiffness in a single direction of the JTC bearing is changed, the overall mode shapes remain similar, and the position of the maximum displacement does not change. The main difference lies in the value of maximum displacement. the following figures provide three order vibration shapes of the crankshaft system as the horizontal/vertical/axial stiffness varies.

Figure 3.44 shows the first-order vibration shapes under different horizontal stiffness. Figure 3.44a, d and g show the first-order vibration shapes under three horizontal

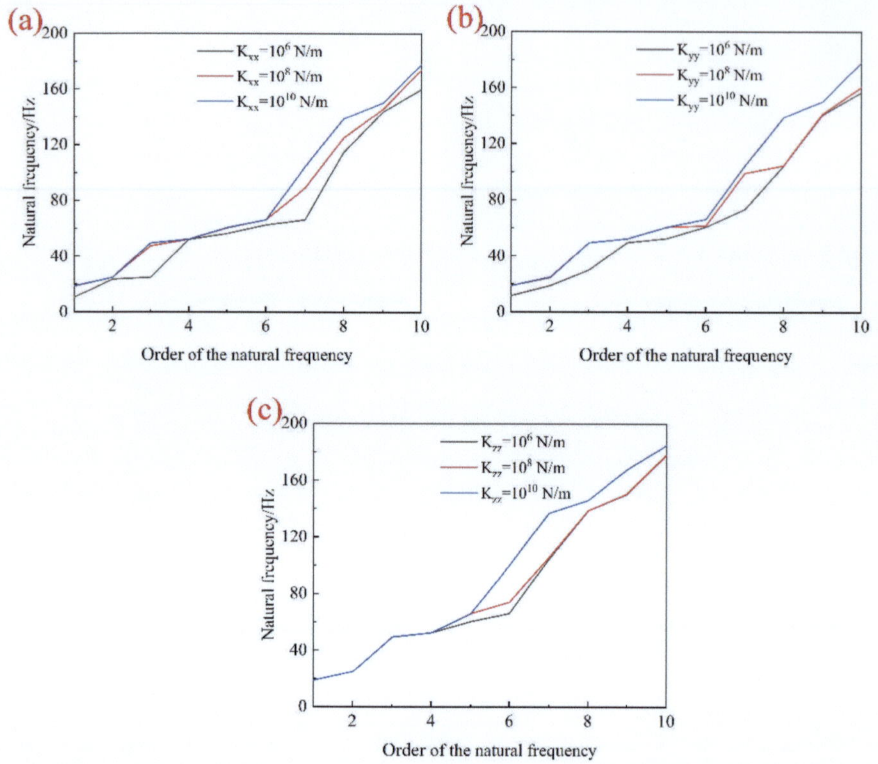

Fig. 3.43 Comparison of natural frequencies under different stiffness: **a** horizontal stiffness; **b** vertical stiffness; **c** axial stiffness

stiffnesses respectively, compared with the results calculated at the horizontal stiffness of 1E8 and 1E10 N/m, the vibration shapes were significantly different when the horizontal stiffness was 1E6 N/m, and MX occurred at the position of JTC bearing.

Similarly, According to Fig. 3.45, it can be found that when the vertical stiffness was 1E6 N/m, MX occurred at the position of the JTC bearing, while it changed to the middle part when the vertical stiffness increases into 1E8 N/m or 1E10 N/m. This is attributed to the larger radial stiffness of JTC bearing compared to the main bearing, exerting a significant influence on the vibration shape. Therefore, it can be speculated that during the operation of the crankshaft system, when sudden axial movement occurs and the JTC bearing is poorly lubricated, the stiffness suddenly increases, which will lead to vibration problems.

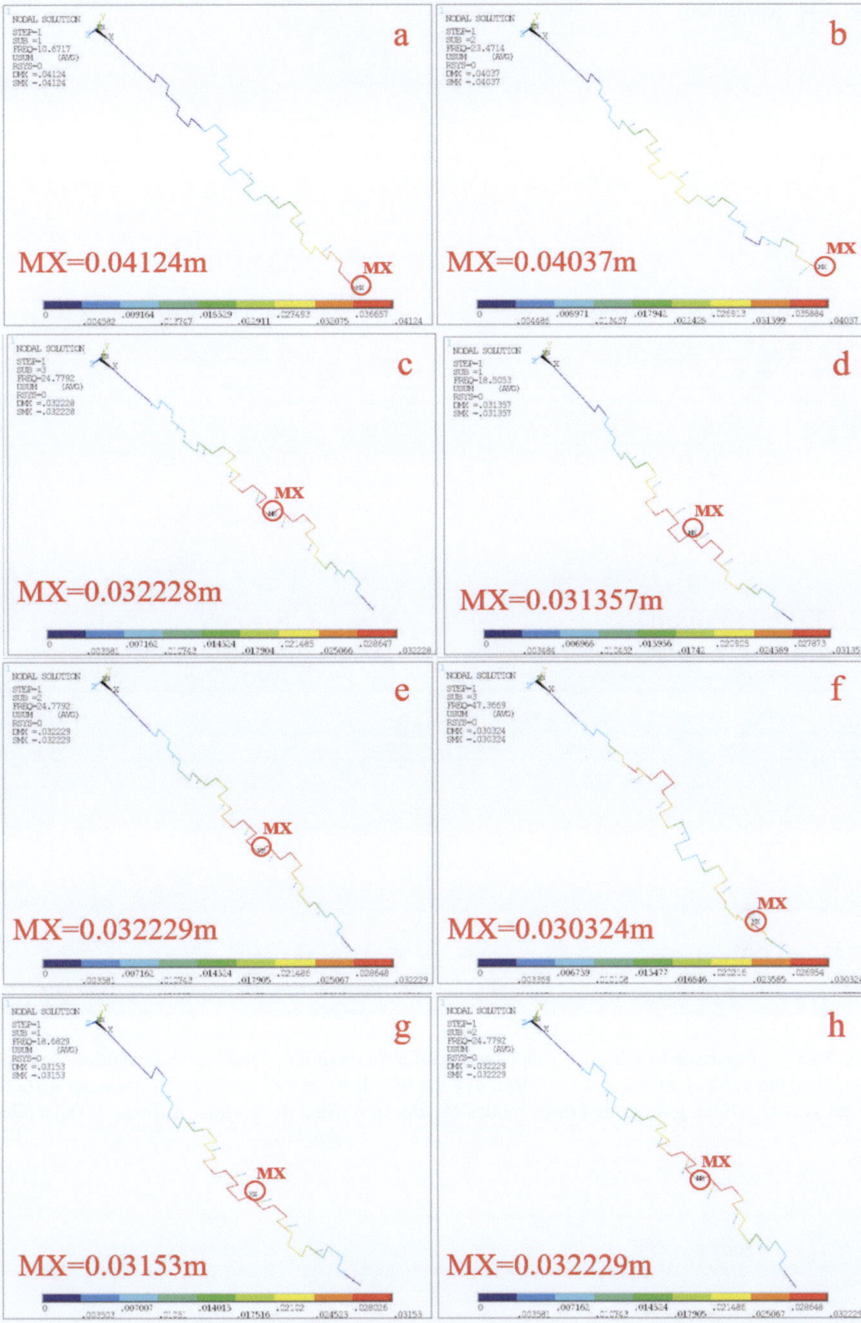

Fig. 3.44 Comparison of vibration shape under different model: **a–c** the first three modes of vibration when the horizontal stiffness is 1E6 N/m; **d–f** the first three modes of vibration when the horizontal stiffness is 1E8 N/m; **g–i** the first three modes of vibration when the horizontal stiffness is 1E10 N/m

Fig. 3.44 (continued)

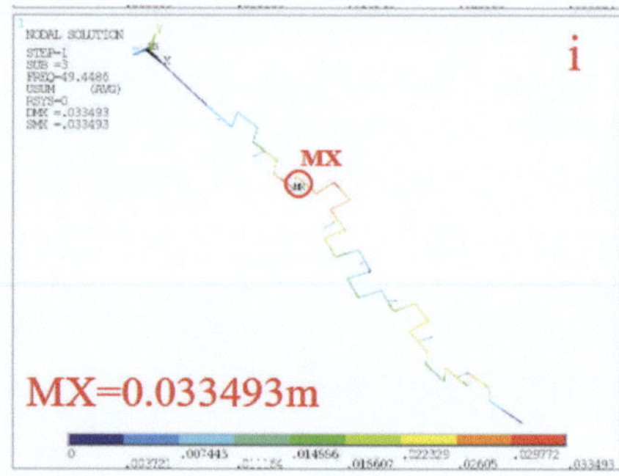

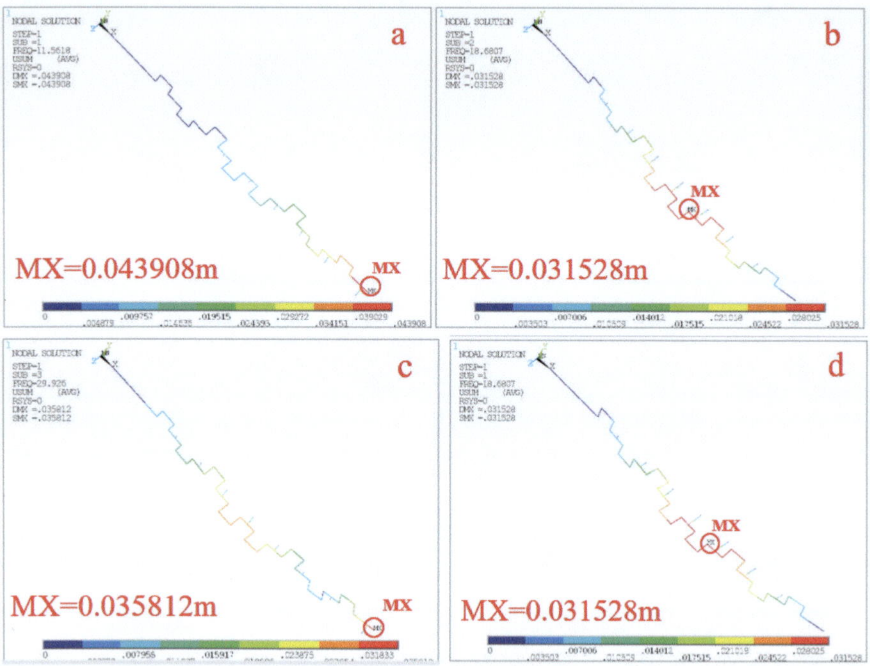

Fig. 3.45 Comparison of vibration shape under different model: **a**–**c** the first three modes of vibration when the vertical stiffness is 1E6 N/m; **d**–**f** the first three modes of vibration when the vertical stiffness is 1E8 N/m; **g**–**i** the first three modes of vibration when the vertical stiffness is 1E10 N/m

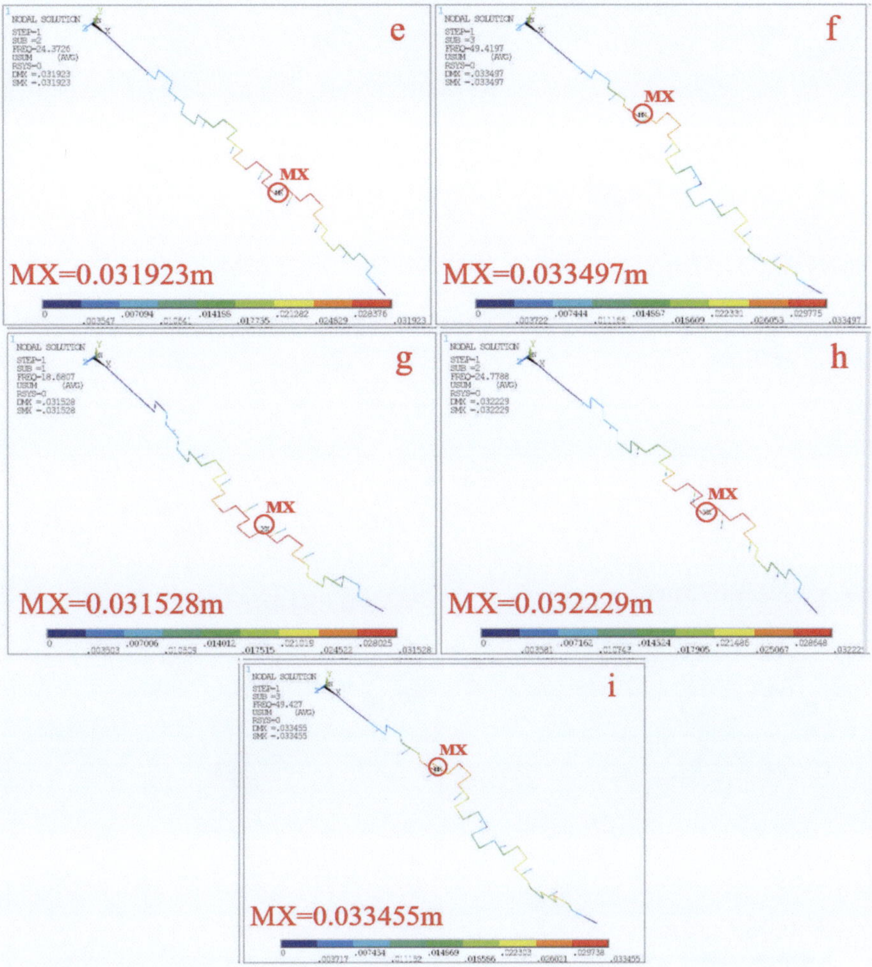

Fig. 3.45 (continued)

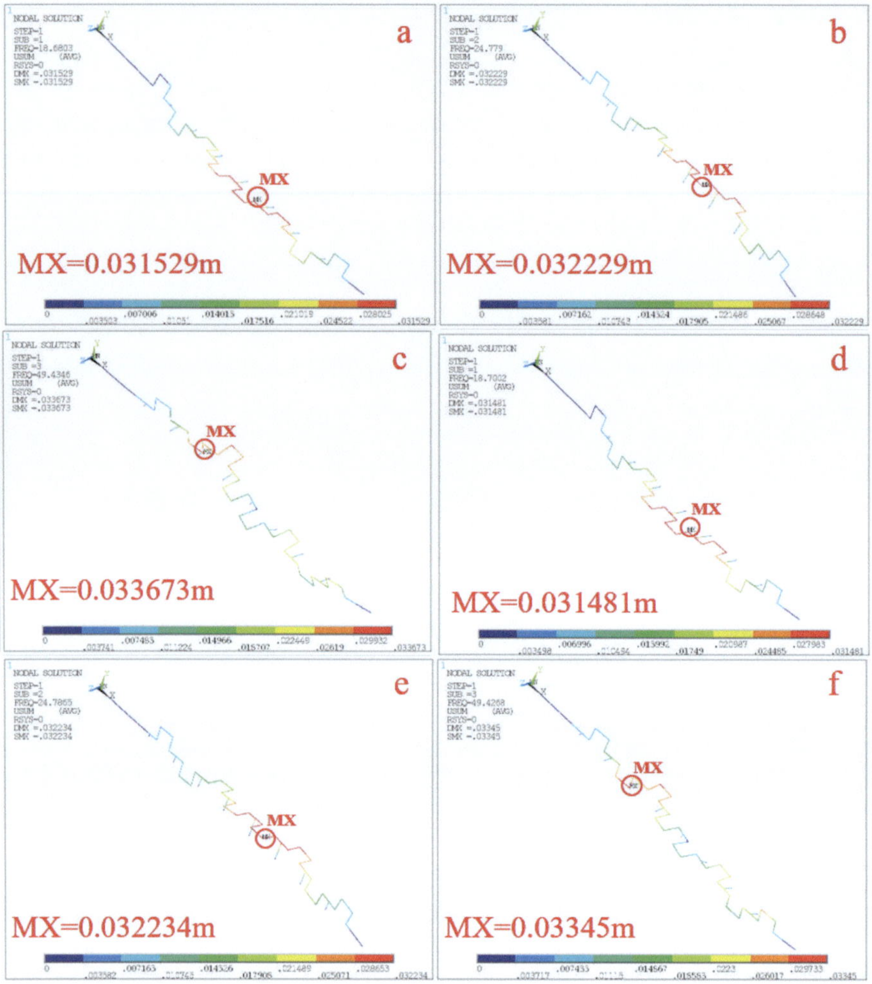

Fig. 3.46 Comparison of vibration shape under different model: **a–c** the first three modes of vibration when the axial stiffness is 1E6 N/m; **d–f** the first three modes of vibration when the axial stiffness is 1E8 N/m; **g–i** the first three modes of vibration when the axial stiffness is 1E10 N/m

Based on Fig. 3.46, the changes in the vibration shape of the same order were relatively small under three axial stiffness inputs. So compared with radial stiffness, axial stiffness has a relatively small impact on the vibration shape.

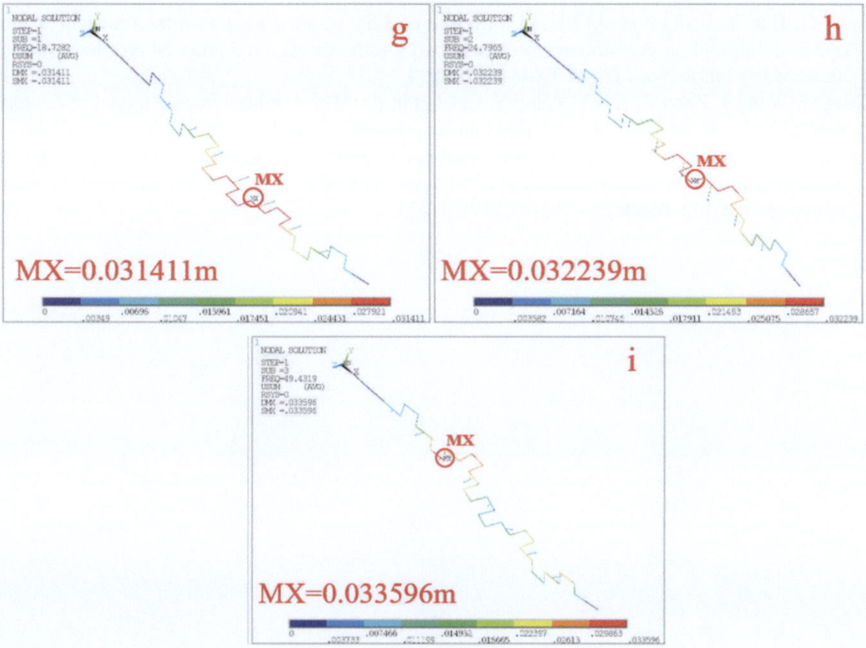

Fig. 3.46 (continued)

References

Arif M, Kango S, Shukla DK (2022) Analysis of textured journal bearing with slip boundary condition and pseudoplastic lubricants. Int J Mech Sci 228:107458

Cao J, Zhai L, Luo Y et al (2022) Transient thermo-elasto-hydrodynamic analysis of a bidirectional thrust bearing in start-up and shutdown processes. Eng Comput 39(4):1511–1533

Li B, Sun J, Zhu S et al (2019) Thermohydrodynamic lubrication analysis of misaligned journal bearing considering the axial movement of journal. Tribol Int 135:397–407

Liang P, Li X, Guo F et al (2022) Influence of sea wave shock on transient start-up performance of water-lubricated bearing. Tribol Int 167:107332

Ma MT, Offner G, Loibnegger B et al (2003) A fast approach to model hydrodynamic behaviour of journal bearings for analysis of crankshaft and engine dynamics. Tribol Ser 43:313–327

Oliveira MVM, Cunha BZ, Daniel GB (2021) A model-based technique to identify lubrication condition of hydrodynamic bearings using the rotor vibrational response. Tribol Int 160:107038

Shi J, Zhao B, He T et al (2023) Tribology and dynamic characteristics of textured journal-thrust coupled bearing considering thermal and pressure coupled effects. Tribol Int 180:108292

Stefani F (2006) THD analysis of a coupled journal and tilting pad thrust bearing system. In: Proceedings of the 5th international tribology conference, AITC-AIT, vol 10

Wang Y, Wang QJ, Lin C (2002) Mixed lubrication of coupled journal-thrust bearing systems. CMES Comput Model Eng. 3(4):517–530

Wang Y, Wang Q, Lin C (2003) Mixed lubrication of coupled journal-thrust-bearing systems including mass conserving cavitation. J Tribol 125:747–755

Xiang G, Han Y, Chen R, Wang J, Ni X (2019) A numerical method to investigate the mixed lubrication performances of journal-thrust coupled bearings. Idn Lubr Tribol 71:1099–1107

Xiang G, Han Y, Wang J et al (2021a) Influence of axial microvibration on the transient hydrodynamic lubrication performance of misaligned journal-thrust microgrooved coupled bearings under water lubrication. Tribol T 64(4):579–592

Xiang G, Wang J, Zhou C, Shi Y, Wang Y, Cai J et al (2021b) A tribo-dynamic model of coupled journal-thrust water-lubricated bearings under propeller disturbance. Tribol Int 160:107008

Xie Z, Shen N, Zhu W et al (2021) Theoretical and experimental investigation on the influences of misalignment on the lubrication performances and lubrication regimes transition of water lubricated bearing. Mech Syst Signal 149:107211

Chapter 4
Optimization Design for Journal-Thrust Couple Bearing Surface Texture Based on Particle Swarm Algorithm

Abstract The journal-thrust coupled bearing (abbreviated as JTC bearing) works at the end of the crankshaft in diesel engine, with the function of carrying radial load and prevent axial movement of the crankshaft. The journal part and the thrust part of this special bearing both form the oil film lubrication when subjected to the time-varying loads in axial and radial directions during the working condition. In this work, the JTC bearing in marine diesel engines was analyzed using a transient thermal elastohydrodynamic (TEHD) lubrication model. The thermal and pressure (thermal-pressure) coupled effects were taken into consideration based on the flow, pressure, and thermal continuity conditions between the journal and thrust parts. Moreover, the transient time-varying load was considered in the lubrication analysis to explore the coupled effect under the actual operating conditions. A test rig was built to validate the established transient TEHD lubrication model for the JTC bearing. The thermal-pressure coupled effect between the journal and thrust parts was found to significantly affect the transient lubrication behaviors of the JTC bearing, which cannot be ignored. Also, different boundary conditions were investigated for the influence of the coupled effect on transient lubrication characteristics, including axial clearances, radial clearances, waveforms of radial load, incline angles and inlet oil temperatures. This chapter could theoretically support the lubrication design of the JTC bearings used in the diesel engine crankshafts.

Keywords Coupled effect · Texture design · Orthogonal test method · Particle swarm algorithm

4.1 Introduction

Bearings, as crucial components in mechanical systems, play a pivotal role in determining the operational efficiency and lifespan of the entire system. Lubrication performance, an important aspect of bearing research, has been the focus of many scholars. To improve the lubrication performance of bearings and extend the service life of mechanical systems, researchers have conducted numerous studies (Xu et al.

B. Zhao et al., *Lubrication and Dynamic Characteristics Analysis Technology*,
https://doi.org/10.1007/978-981-95-2270-5_4

2022; Chen et al. 2024). Among these, arranging micro-textures on the bearing surface has been considered an effective method to enhance lubrication performance. Since Hamilton et al. (1966) first proposed that the distribution of microstructures on the surfaces of mechanical friction pairs can significantly enhance their hydrodynamic lubrication state, researchers have extensively studied the effects of textures on bearing lubrication. These studies have addressed many aspects, including the design and fabrication techniques of bearing surface textures (Ibatan et al. 2015; Allen and Raeymaekers 2021), the mechanism by which textures improve lubrication, and the effects of texture characteristic parameters on lubrication performance (Guo et al. 2021; Long et al. 2024).

So far, extensive research has been conducted on lubrication with surface textures for different kinds of bearings. For journal bearings, Galda et al. (2019) analyzed the characteristic parameters during the transition of lubrication regimes and compared the lubrication performance of textured and non-textured bearings. For thrust bearings, Wu et al. (2023) developed three types of textures, conducted frequency domain analysis, and found that gradient groove textures can effectively improve lubrication conditions. In addition to these conventional bearings, research on the lubrication performance of some special types of textured bearings has also been carried out. Pei et al. (2016) used a deterministic lumped parameter thermal model to study the impact of nine textures on the lubrication performance of floating ring bearings, guiding the development of turbochargers. Xie et al. (2024) studied the internal flow field characteristics and evolution rules of the surface texture of a new type of water-lubricated bearing and explored the mechanism of the texture's influence on hydrodynamic lubrication.

Moreover, related studies have found that the structural parameters of textures directly affect their optimization performance. Therefore, the specific design of texture parameters (Tala-Ighil et al. 2015; Hu et al. 2019) is also a hot topic in bearing texture research. The most common method for surface texture design is trial and error, where researchers conduct parameterized analysis on texture structure parameters such as shape and width through assumed methods (Gropper et al. 2016). A properly designed texture enhances the micro-hydrodynamic effect of the lubricant, delaying oil film rupture and improving lubrication performance. However, such methods still require pre-setting texture optimization parameter combinations, with each parameter setting having a certain degree of arbitrariness, leading to the possibility that the true optimal solution may not be included in the preset options (Mavrovouniotis et al. 2017). Moreover, the optimization efficiency is relatively low. Therefore, based on progressively improving lubrication analysis models, developing efficient parameter optimization methods is a key focus for future texture design (Codrignani et al. 2020; Guo et al. 2023). To enhance the efficiency of texture optimization, researchers have attempted to use intelligent algorithms to optimize the lubrication design of radial bearings or thrust bearings, such as the domain decomposition method (Pei et al. 2011), sequential quadratic programming (SQP) method (Wang et al. 2020), genetic algorithm (GA) (Chalkiopoulos et al. 2020; Zhang et al. 2018), particle swarm optimization (PSO) algorithm (Chan 2015; Zhang et al. 2021), and others. Studies have shown that intelligent algorithms can significantly improve the speed and accuracy of texture optimization.

The journal-thrust coupled (JTC) bearing, as a specialized bearing at the end of a diesel engine crankshaft system, comprises both journal and thrust parts. Unlike conventional bearings, JTC bearings require consideration of both journal and thrust part lubrication when designing surface textures. Considering the complexity of the operating environment, there is a dual requirement: on one hand, reducing the friction coefficient of the journal part, while on the other hand, enhancing the load-carrying capacity of the thrust part. Currently, research on this type of bearing is limited. Since Wang et al. (2018) first proposed that the lubrication model should consider the continuity conditions between the journal part and thrust part, the lubrication model of JTC bearings has gradually evolved from hydrodynamic (HD) to elastohydrodynamic (EHD) and even thermo-elastohydrodynamic (TEHD) models (Xiang et al. 2021; Shi et al. 2023, 2024). Based on this foundation, researchers have recently begun to study the surface texture of these bearings. Xiang et al. (2020) conducted a study on JTC bearings with groove texture surfaces, investigating the effects of three different bottom texture shapes on bearing lubrication performance and exploring their lubrication mechanisms. However, the design of texture parameters in this study is preliminary and mainly relies on assumptions, which may not necessarily result in texture parameters that maximize the lubrication effect.

This study specifically focused on JTC bearings and, by considering the continuity conditions of both the journal and thrust parts, established a lubrication model that accounted for the coupled effects. Based on this model, the texture parameters with the greatest impact on lubrication performance were determined through orthogonal tests. Subsequently, these parameters were optimized using a particle swarm optimization algorithm. This formed a complete process for optimizing texture parameters, guiding the design of surface textures for JTC bearings.

Nomenclature

h	Oil film thickness	ω	Relative speed of journal bearing
c	Bearing clearance	ρ	Lubricant density
e	Eccentricity	η_0	Lubricant viscosity at supply temperature
R_2	Outer radius of thrust part	β	Viscosity temperature coefficient
R_1	Inner radius of thrust part	T	Film temperature
θ	Position angle	T_0	Supply temperature
φ	Attitude angle	h_b	HEAT transfer coefficient
γ	Inclination angle	α	Thermal expansivity of lubricant
Y	Axial position	U_θ	Circumferential speed of the lubricant in thrust part

BL	Axial length of the bearing	U_r	Radial speed of the lubricant in thrust part
Ω	Calculation area	U_J	Linear speed of journal part
h_0	Geometric film thickness in thrust part	u_J	Radial speed of the lubricant in journal part
r	Radial coordinates	v_J	Circumferential speed of the lubricant in journal part
η	Lubricating oil viscosity	w_J	Axial speed of the lubricant in journal part
Nr	Rotational speed	**Subscript**	
k	Thermal conductivity	x	x (horizontal) direction
h_h	Convection heat transfer coefficient	y	y (vertical) direction
ϕ_s	Shear flow factor	J	Journal part
ϕ_θ	Circumferential pressure-flow factors	T	Thrust part
ϕ_z	Axial pressure-flow factors	JE	Elastic deformation in journal part
ϕ_r	Radial pressure-flow factors	JT	Thermal deformation in journal part
ϕ_c	Contact factor	TE	Elastic deformation in thrust part
R_0	Inner diameter of bearing	TT	Thermal deformation in thrust part
P	Film pressure	l	Lubricant
ε	Eccentricity ratio	b	Bearing
		j	Journal

4.2 Simulation Model

Based on the lubrication model of the journal-thrust coupled bearing, this work used the PSO algorithm to optimize the design of the bearing surface texture. The schematic diagrams of the journal part and the thrust part are shown in Fig. 4.1, and the specific model is as follows.

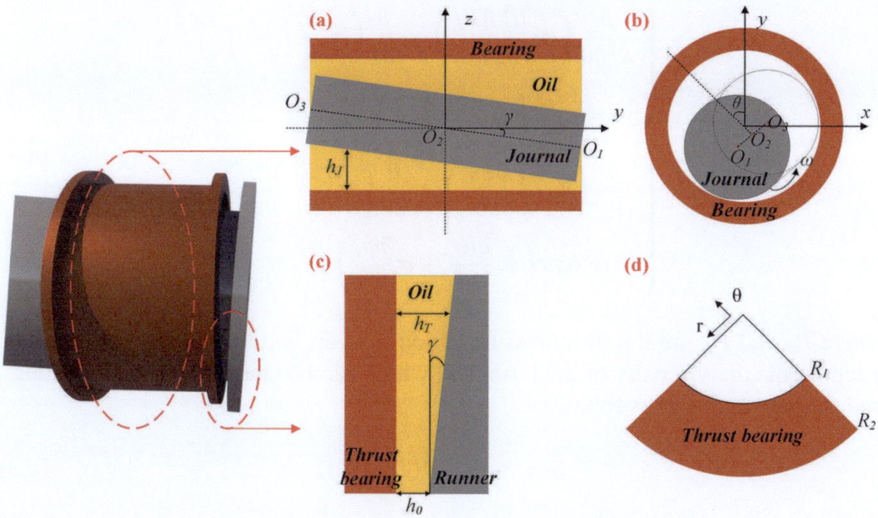

Fig. 4.1 The schematic diagrams of the journal part and the thrust part

4.2.1 Lubrication Model

Firstly, the film thickness equation was established based on the geometry of the bearing. Due to the tilt of the journal, O_1, O_2, and O_3 were used to represent the center positions of the front, middle, and rear sections of the journal, respectively. The calculations for the two parts are as follows:

$$\begin{cases} h_J = c + \mathrm{e}\cos(\theta - \varphi) + \tan\gamma\cos(\theta - \varphi)\left(Y - \dfrac{BL}{2}\right) + h_{JE} + h_{JT} + h_{TexJ} \\ h_T = h_0 + r\sin\gamma\cos\theta + h_{TE} + h_{TT} + h_{TexT} \end{cases} \tag{4.1}$$

where h_J and h_T are the film thickness of journal part and thrust part, respectively, and h_{TexJ} and h_{TexT} are the height of the texture in the journal part and thrust part, respectively. The elastic deformation and thermal deformation of the bearing are calculated based on the influence coefficient method.

The Reynold equations for the two parts are as follows:

$$\begin{cases} \frac{\partial}{R_0^2 \partial\theta}\left(\phi_\theta \frac{h_J^3}{\eta}\frac{\partial P_J}{\partial\theta}\right) + \frac{\partial}{\partial z}\left(\phi_z \frac{h_J^3}{\eta}\frac{\partial P_J}{\partial z}\right) \\ \quad = \frac{6U_J}{R_0}\left(\phi_c \frac{\partial h_J}{\partial\theta} + \sigma\frac{\partial\varphi_s}{\partial\theta}\right) \\ \frac{\partial}{r\partial\theta}\left(\phi_\theta \frac{h_T^3}{\eta}\frac{\partial P_T}{\partial\theta}\right) + \frac{\partial}{\partial r}\left(\phi_r \frac{rh_T^3}{\eta}\frac{\partial P_T}{\partial r}\right) \\ \quad = 6\omega r\left(\phi_c \frac{\partial h_T}{\partial\theta} + \sigma\frac{\partial\phi_s}{\partial\theta}\right) \end{cases} \tag{4.2}$$

where P_J and P_T are the film pressure of journal part and thrust part, respectively. η represents the viscosity of the lubricant, which can be changed with temperature, and the calculation is as follows:

$$\eta = \eta_0 \exp(\beta(T - T_0)) \tag{4.3}$$

The lubrication model considered the thermal effect of the bearing and the lubricant, and calculated the temperature of the lubricant and bearing to obtain the thermal deformation and the viscosity of the lubricant. The film temperature of journal part T_J and thrust part T_T can be calculated based on the following equation:

$$\begin{cases} \rho C_p\left(\frac{v_J}{r}\frac{\partial T_J}{\partial\theta} + w_J\frac{\partial T_J}{\partial z}\right) = \alpha T_J\left(w_J\frac{\partial P_J}{\partial z} + v_J\frac{\partial P_J}{r\partial\theta}\right) \\ \quad + k_l\frac{\partial^2 T_J}{\partial r^2} + \eta\left[\left(\frac{\partial u_J}{\partial z}\right)^2 + \left(\frac{\partial v_J}{\partial z}\right)^2\right] \\ \rho C_p\left(U_r\frac{\partial T_T}{\partial r} + \frac{U_\theta}{r}\frac{\partial T_T}{\partial\theta}\right) = \alpha T_T\left(U_r\frac{\partial P_T}{\partial r} + U_\theta\frac{\partial P_T}{r\partial\theta}\right) \\ \quad + k_l\frac{\partial^2 T_T}{\partial z^2} + \eta\left[\left(\frac{\partial U_r}{\partial z}\right)^2 + \left(\frac{\partial U_\theta}{\partial z}\right)^2\right] \end{cases} \tag{4.4}$$

For the bearing temperature of the two parts, the following equation based on heat conduction can be used to calculate:

$$\begin{cases} \frac{\partial^2 T_J}{\partial x^2} + \frac{\partial^2 T_J}{\partial y^2} + \frac{\partial^2 T_J}{\partial z^2} = 0 & \textit{journal part} \\ \frac{\partial^2 T_T}{\partial r^2} + \frac{1}{r}\frac{\partial T_T}{\partial r} + \frac{1}{r^2}\frac{\partial^2 T_T}{\partial\theta^2} + \frac{\partial^2 T_T}{\partial z^2} = 0 & \textit{thrust part} \end{cases} \tag{4.5}$$

The boundary conditions used in the thermal analysis of the journal part and the thrust part are as follows, where n is the unit vector in the convection heat transfer direction:

$$\begin{cases} k_b \frac{\partial T_J}{\partial n} = -h_b(T_J - T_\infty) & \textit{bearing-environment} \\ k_l \frac{\partial T_J}{\partial r} = k_b \frac{\partial T_J}{\partial r} & \textit{bearing-lubricant} \\ k_J \frac{\partial T_J}{\partial r} = \frac{1}{2\pi} k_l \int_0^{2\pi} \frac{\partial T_J}{\partial r} d\theta & \textit{journal-lubricant} \end{cases} \tag{4.6}$$

$$\begin{cases} k_b \frac{\partial T_T}{\partial n} = -h_b(T_T - T_\infty) & \textit{bearing-environment} \\ k_l \frac{\partial T_T}{\partial z} = k_b \frac{\partial T_T}{\partial z} & \textit{bearing-lubricant} \\ k_J \frac{\partial T_T}{\partial z} = \frac{1}{2\pi} k_l \int_0^{2\pi} \frac{\partial T_T}{\partial z} d\theta & \textit{journal-lubricant} \end{cases} \tag{4.7}$$

In contrast to conventional lubrication models, the 'coupled model' established in this work considered the continuity conditions of the lubricant between the journal and thrust parts. Specifically, these conditions include the pressure continuity condition, the flow continuity condition, and the heat flow continuity condition at the journal-thrust interface, as shown in Fig. 4.2. The expressions for these conditions are as follows:

$$\begin{cases} (P_J)_{j,m} = (P_T)_{j,m} & \textit{pressure continuity} \\ \left(\frac{h_J^3}{12\eta} \frac{\partial P_J}{\partial z}\right)_{j,m} = -\left(\frac{h_T^3}{12\eta} \frac{\partial P_T}{\partial r}\right)_{j,m} & \textit{flow continuity} \\ \left(w_J \frac{\partial T_J}{\partial z}\right)_{j,m} = -\left(U_r \frac{\partial T_T}{\partial r}\right)_{j,m} & \textit{heat flow continuity} \end{cases} \tag{4.8}$$

where, the specific expression of the velocity w_J and U_r are as follows, and $d = R_0$-r represents distance from bearing surface to selected point:

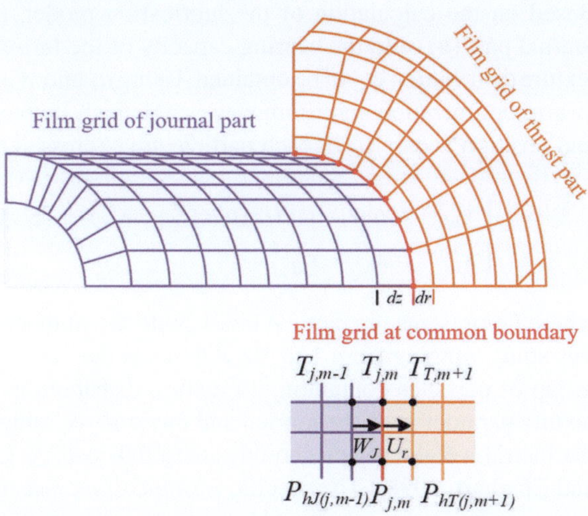

Fig. 4.2 Schematic diagram of flow, pressure and thermal continuity conditions

$$\begin{cases} w_J = \dfrac{h^3}{h_T^3}\phi_z \dfrac{1}{2\eta}\dfrac{\partial p_J}{\partial z}\left(d^2 - dh_T\right) \\ U_r = \dfrac{\partial p_T}{\partial r}\left(\displaystyle\int_0^z \frac{z\mathrm{d}z}{\eta} - \frac{\int_0^{h_{\text{thrust}}} \frac{\mathrm{d}z}{\eta}}{\int_0^{h_{\text{thrust}}} \frac{z\mathrm{d}z}{\eta}} \int_0^z \frac{\mathrm{d}z}{\eta}\right) \end{cases} \tag{4.9}$$

Based on the calculated film pressure of each part, the friction coefficient of the journal part μ_J and the bearing capacity of the thrust part W_{th} can be obtained as follows:

$$\mu_J = \frac{\int\int\left(\frac{\eta U_J}{h_J} + \frac{h_J}{2}\frac{\partial P_J}{\partial x}\right)R_0\mathrm{d}\theta\mathrm{d}z}{\int\int P_J R_0 \mathrm{d}\theta\mathrm{d}z} \tag{4.10}$$

$$W_{th} = \int\int P_T r\mathrm{d}\theta\mathrm{d}r \tag{4.11}$$

4.2.2 *Particle Swarm Optimization Algorithm Model*

Based on the calculation of the lubrication model, the friction coefficient of the journal part (μ_J) and the bearing capacity of the thrust part (W_{th}) under the specific texture distribution could be obtained. Using μ_J and W_{th} as fitness functions, a particle swarm optimization algorithm was established. In the model, the velocity vector v_{id} and position vector x_{id} of each particle are expressed as follows:

$$\begin{cases} \boldsymbol{v}_{id}^{T+1} = \omega_i \boldsymbol{v}_{id}^{T} + c_1 r_1\left(pbest_i^T - \boldsymbol{x}_{id}^T\right) + c_2 r_2\left(gbest^T - \boldsymbol{x}_{id}^T\right) \\ \boldsymbol{x}_{id}^{T+1} = \boldsymbol{x}_{id}^{T} + \boldsymbol{v}_{id}^{T+1} \end{cases} \tag{4.12}$$

where i represents the particle index, with the number of particle swarm set to 30 in this study, ranging from 1 to 30; d denotes the particle dimension index, with each group of particles containing 20 texture distributions corresponding to the groove texture parameters of the journal and thrust parts, ranging from 1 to 20; ω_i stands for the inertia weight, which should satisfy $0.9 \le \omega_i \le 1.2$ according to the study (Shi and Eberhart, 1998). The specific formula of ω_i is as follows:

$$\omega_i = \omega_{\max} - \frac{(\omega_{\max} - \omega_{\min}) \times T}{T_{\max}} \tag{4.13}$$

where $\omega_{\max}$ and $\omega_{\min}$ represent the maximum and minimum inertia weights during the population evolution process, with $\omega_{max} = 0.9$ and $\omega_{max} = 0.4$; T denotes the evolution generation; $T_{\max}$ indicates the maximum evolution generation, with $T_{\max} = 60$. c_1 and c_2 are referred to as learning factors or acceleration coefficients, with $c_1 = c_2 = 2$; r_1 and r_2 are random numbers between 0 and 1; $pbest_i{}^T$ denotes the individual best solution; $gbest^T$ represents the global best solution.

The velocity vector and position vector boundary conditions of the particle swarm in the calculation are:

$$\begin{cases} v(v \geq v_{\max}) = v_{\max} = 4 \\ v(v \leq v_{\min}) = v_{\min} = -4 \\ x(x \geq x_{\max}) = x_{\max} = BP \\ x(x \leq x_{\min}) = x_{\min} = 0 \end{cases} \tag{4.14}$$

where *BP* is the boundary parameter of the bearing. Taking the bearing length as an example, for the journal part, *BP* is equal to the axial length of the bearing, while for the thrust part, *BP* is equal to the radial length of the bearing.

4.2.3 Simulation Procedure

Figure 4.3 shows the flow chart of the PSO algorithm used to calculate optimal texture parameters. The entire calculation process can be divided into the lubrication module and the optimization algorithm module. The lubrication module is more complex compared to conventional bearing lubrication models, as it requires consideration of continuity conditions. After separately calculating the film pressure of the journal part and the thrust part, the flow continuity condition is used to establish connections between each part and modify the film pressure accordingly. Similarly, the thermal continuity condition is used to couple the relationship of the film temperature of each part after temperature calculation. As for the optimization algorithm module, it serves as the outer loop of the lubrication calculation. The texture parameters corresponding to each particle swarm are substituted into the lubrication program for calculation. After completing all lubrication calculations, the friction coefficient of the journal part and the bearing capacity of the thrust part under each texture distribution are obtained. Based on this, the texture quality is evaluated, and the partial optimal and global optimal texture distributions are determined. These optimal distributions are then substituted into Eq. (4.12) for the next calculation. After looping to the set evolutionary algebra, the best optimization result can be obtained and outputted.

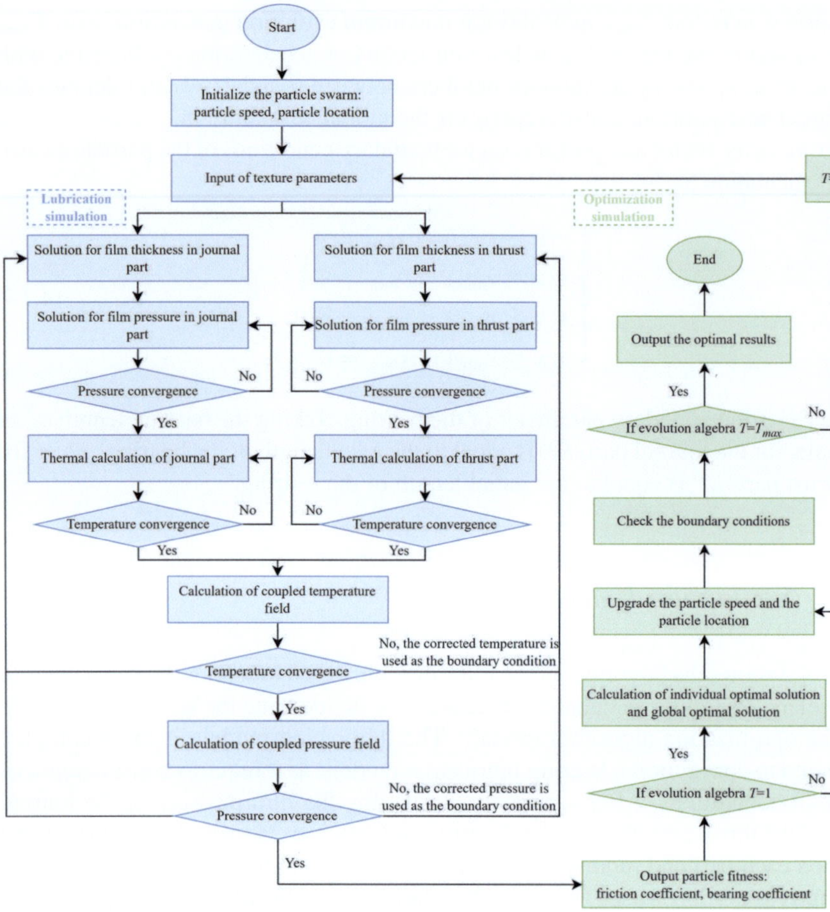

Fig. 4.3 Flow chart of the texture optimization using PSO algorithm

4.3 Results and Discussion

4.3.1 Model Verification

Before conducting PSO calculations, validation of the lubrication model has been performed through experimentation using a constructed test rig (Shi et al. 2023). Figure 4.4 shows the photograph and the schematic diagram of the test rig. Two resistance temperature sensors (PT100) were arranged in the journal part and the thrust part respectively to measure the bearing temperature, and the friction force was measured by the torque sensor to calculate the friction coefficient. The relevant experimental parameters are shown in Table 4.1.

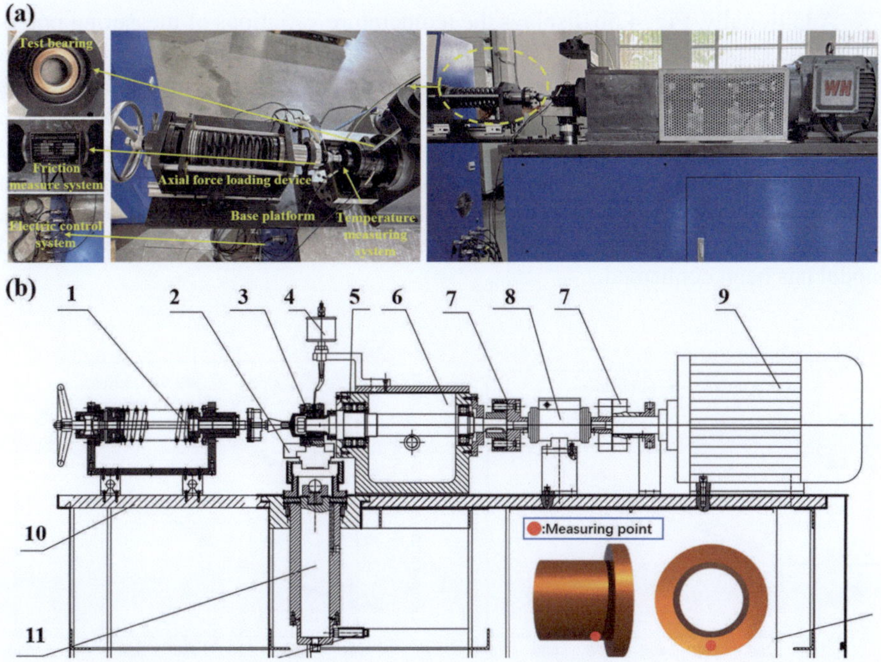

1- Axial force loading device; 2- Oil receiving box; 3- Test fixture; 4- Oil box; 5-Test bearing; 6-Bearing block; 7- Coupling; 8- Torque sensor; 9-Motor; 10-Base platform; 11- Radial force loading device

Fig. 4.4 Photograph and schematic diagram of the test rig; **a** photograph; **b** schematic diagram

Table 4.1 Experiment parameters

Parameters	Value
Internal diameter of journal part	40 mm
External diameter of thrust part	60 mm
Inclination angle	0.002°
Velocity	500 r/min
Viscosity of the oil	0.175 Pa s
Supply temperature	20 °C
Radial load	5500 N
Vertical load	100/300/500/700 N

The comparison between simulation results and experimental results are shown in the Fig. 4.5. As depicted in Fig. 4.5a, the friction coefficients obtained from both experiment and simulation under different axial loads exhibit a close correspondence. The deviations between the test results and the simulation results under each load were less than 5%. The maximum deviation occurred at a load of 500 N, where the test result was 0.0087 and the simulation result was 0.00856, resulting in a deviation of

2%. Additionally, Fig. 4.5b displays the temperature variations of measuring points during the experiment under a 500 N axial load, with the stabilized value after 2 h of operation considered as the test result. Correspondingly, Fig. 4.5c demonstrates that the simulated temperature predictions at each measuring point align well with the experimental observations. Similarly, the deviations between the test results and simulation results for temperature were also less than 5%. The maximum deviation in the journal part was 2%, and the maximum deviation in the thrust part was 3%, both of which are within a reasonable range. Thus, the reliability of the established model has been confirmed.

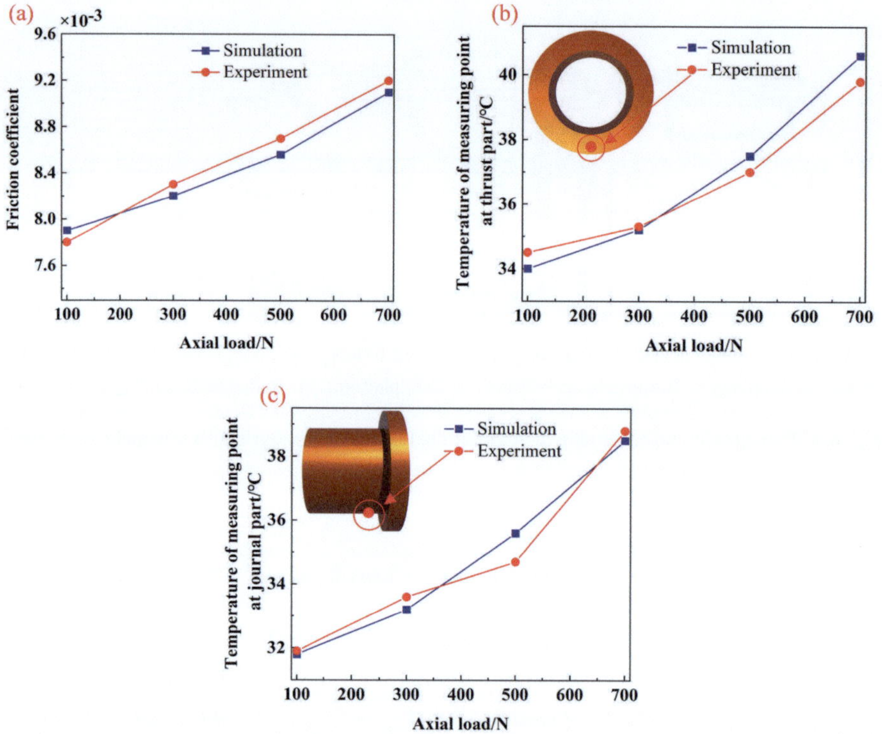

Fig. 4.5 **a** Verification of the friction coefficient; **b** verification of the temperature at the thrust part; **c** verification of the temperature at the journal part

4.3.2 Determination of Optimization Parameters Used in PSO Algorithm

4.3.2.1 Study on the Influence of Different Texture Parameters on Lubrication Performance

The optimization of all texture parameters for improving the lubrication of JTC bearings is impractical due to the multitude of parameters involved. Therefore, it is necessary to select the parameters that have the greatest impact on lubrication performance for subsequent optimization using the PSO algorithm. This work mainly selected optimization objects among texture parameters including length (axial length for the journal part and radial length for the thrust part), width (TW), depth (TD) and spacing (TS). Since the boundary values differ between the journal part and the thrust part, the length ratio (TL) was used to characterize the texture length in each part during calculations, thereby reducing additional boundary condition settings and improving calculation efficiency. Apart from that, the shape of the texture is groove type, and the number of textures is 10 (10 textures were distributed in each journal part and thrust part respectively). The load is chosen as the maximum value of the transient operating condition of the actual machine to simulate harsh real-world conditions, with the direction vertically downward. The texture distribution is primarily concentrated in the main load-bearing area, symmetrically distributed along the circumference of the bearing with 180° symmetry. Table 4.2 shows the parameters used in this work, and Fig. 4.6 depicts the schematic diagram of texture design.

Preliminary parametric analysis was conducted by assuming texture parameter values to study the influence of different texture parameters on the lubrication characteristics of the JTC bearing (μ_J and W_{th}), thus guiding the setting of the texture parameters. The study also incorporated the calculation results of the model without considering coupled lubrication effects (original model) for comparison, thereby illustrating the necessity of considering continuity conditions. The specific parameters are listed in Table 4.2.

The lubrication performance of the JTC bearing (μ_J and W_{th}) under five texture depths T_D can be found in Fig. 4.7, where the 'coupled model' represents the model considered the pressure, flow and heat flow continuity conditions. As the depth of the texture increases, the change trends calculated by the two models are consistent, that is, W_{th} increases and μ_J decreases. In addition, μ_J and W_{th} calculated using the coupled model were both higher than those of the original model. This is because after considering the coupled lubrication effect, the maximum film pressure in the thrust part increases and the W_{th} increases. At the same time, the gradient of journal part pressure in the circumferential direction increases. According to the equation of friction force and friction coefficient, the friction force increases while the bearing capacity remains unchanged, which is equal to the load, resulting in an increase in the friction coefficient. According to the results, within the given interval, the texture depth should be designed to be relatively large.

Table 4.2 Parameters used for calculation

Parameter	Value
Radius clearance/c	200 μm
Axial clearance/h_0	70 μm
Axial load/W_y	700 kN
Journal axial length/BL	0.19 m
Incline angle of journal/γ	0.01°
Supply temperature/T_0	40 °C
Brand of the oil/η_0	15W40
Elastic modulus/E_J	105 GPa
Poisson ratio/v_p	0.327
Velocity/ω	500 r/min
Viscosity temperature coefficient/β	– 0.0491
Thermal expansion coefficient/k_p	1.2E–5/K
Thermal conductivity /k_b	47 W/mK
Heat transfer coefficient /h_b	80 W/m^2 K
Thermal conductivity/k_r	0.13 W/mK
Specific heat capacity/Cp_o	2000 W/m^2 K
Texture depth/T_D	4/6/8/10/12 μm
Texture length ratio/T_L	0.2/0.4/0.6/0.8/1
Texture spacing/T_S	0.015/0.03/0.045/0.06/0.075 m
Width of the texture/ T_W	0.015/0.03/0.045/0.06/0.075 m

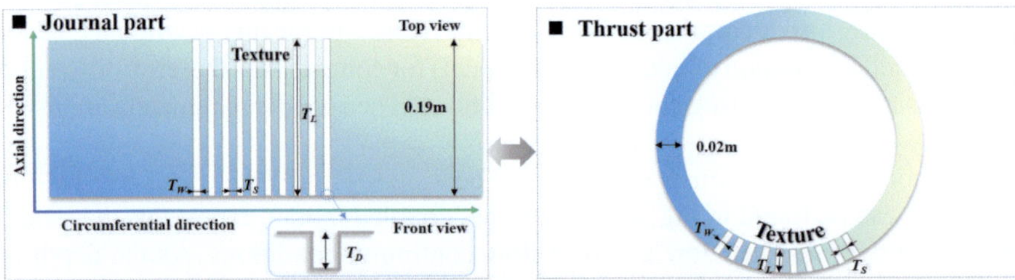

Fig. 4.6 Schematic diagram of texture design

The lubrication performance of the JTC bearing under five texture length ratios T_L can be found in Fig. 4.8. In this section, the texture lengths of the journal and thrust parts are designed simultaneously. Since the maximum texture length of each part is different, the texture length of each part is characterized by the length ratio (0.2/0.4/0.6/0.8/1). It can be observed that both μ_J and W_{th} exhibit non-linear variations with the length ratio. Specifically, μ_J decreases with increasing length ratio within the range of 0 to 0.8 under both models. This suggests that enhancing the texture length

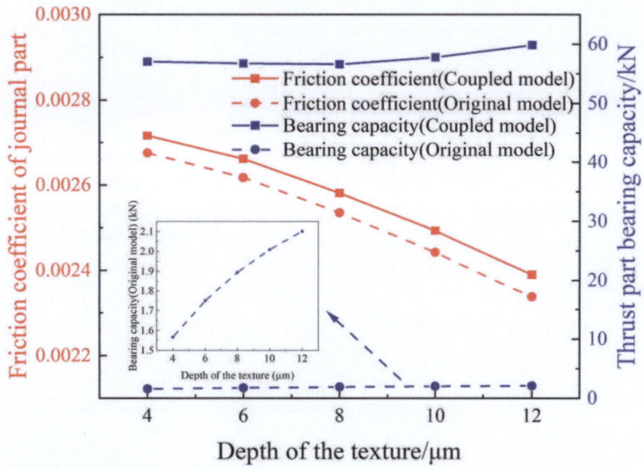

Fig. 4.7 Lubrication performance of the JTC bearing under different texture depths

can initially improve lubrication performance, yet further increases in length may lead to a deterioration in performance. Conversely, W_{th} shows a similar decreasing trend with the length ratio under the original model, consistent with μ_J. However, for the results calculated with the coupled model, W_{th} demonstrates an approximately linear relationship with the texture length. This phenomenon is attributed to the coupled effect, which acknowledges that textures on the journal and thrust parts influence other components as well.

The lubrication performance of the JTC bearing under five texture spacings T_S can be found in Fig. 4.9. As the distance changes, the results calculated by the two models tend to be more consistent. It is observed that with a decrease in the texture distance, there is a corresponding decrease in μ_J. Also, the reduction of μ_J is more pronounced at smaller distance. Conversely, when the texture distance is large, the μ_J exhibits minimal change. This phenomenon can be attributed to the distance being significantly larger than the width of the texture, causing the bearing surface to transition to a smoother profile. Consequently, the efficacy of the texture in enhancing lubrication diminishes. For the thrust part, the W_{th} is relatively low at small distance, which conflicts with the μ_J optimization strategy and requires a trade-off.

The lubrication performance of the JTC bearing under five widths of the texture T_W can be found in Fig. 4.10. Similar to the impact of spacing on lubrication, each μ_J calculated by the two models increases with the rise of T_W. However, for the W_{th}, the results indicate that at smaller widths, W_{th} increases with the widening of the texture width. However, there exists a critical value, beyond which W_{th} decreases with the increase in width. The critical value obtained from the calculations of the two models differs slightly.

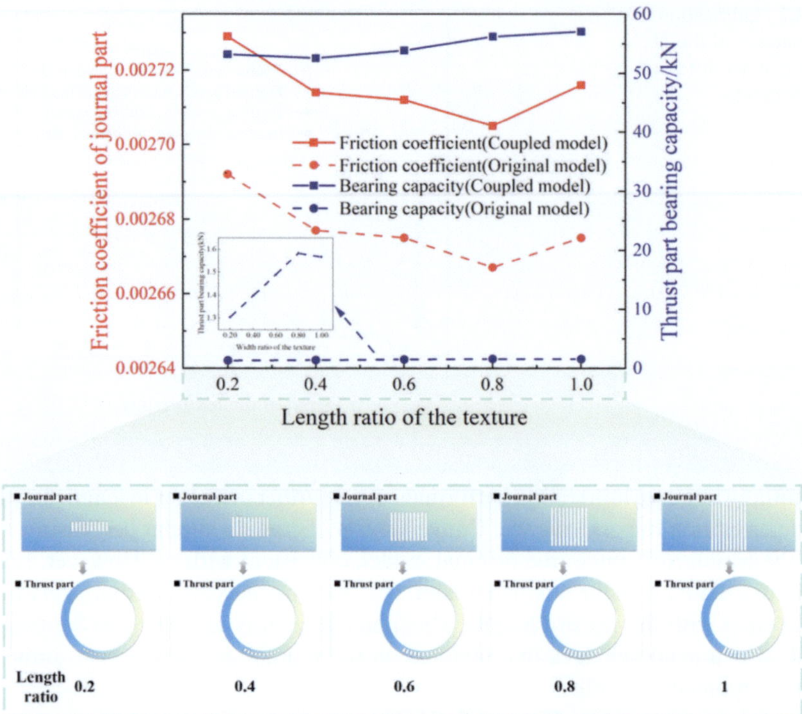

Fig. 4.8 Lubrication performance of the JTC bearing under different texture length ratios

4.3.2.2 Selection of the Optimized Texture Parameters Based on the Orthogonal Test Method

After the previous analysis, it is evident that each texture parameter affects the lubrication performance of JTC bearing differently, and achieving synchronous optimizations for all parameters is impractical. Therefore, it is necessary to identify the parameter with the greatest impact on lubrication performance among the four parameters for subsequent PSO algorithm optimization. The L_{25} orthogonal table was used to design the texture parameters including the depth T_D, length ratio T_L, spacing T_S and width T_W of the texture, and each factor contains five levels, as shown in Table 4.3. The purpose of the calculation is to obtain texture parameters that can improve the W_{th} and reduce the μ_J as much as possible.

The specific orthogonal test plan is shown in the Table 4.4, and the lubrication performance results are shown in Fig. 4.11. It can be found from Fig. 4.11 that among all the test schemes, T_{18} had the smallest μ_J, and T_{25} has the largest W_{th}.

In order to further select one texture parameter that have the greatest impact on the lubrication performance, range analysis was performed based on the orthogonal test results. The range (R_A) represents the importance of each factor on the experimental indicator. A larger R_A indicates that the factor has a greater impact on the target

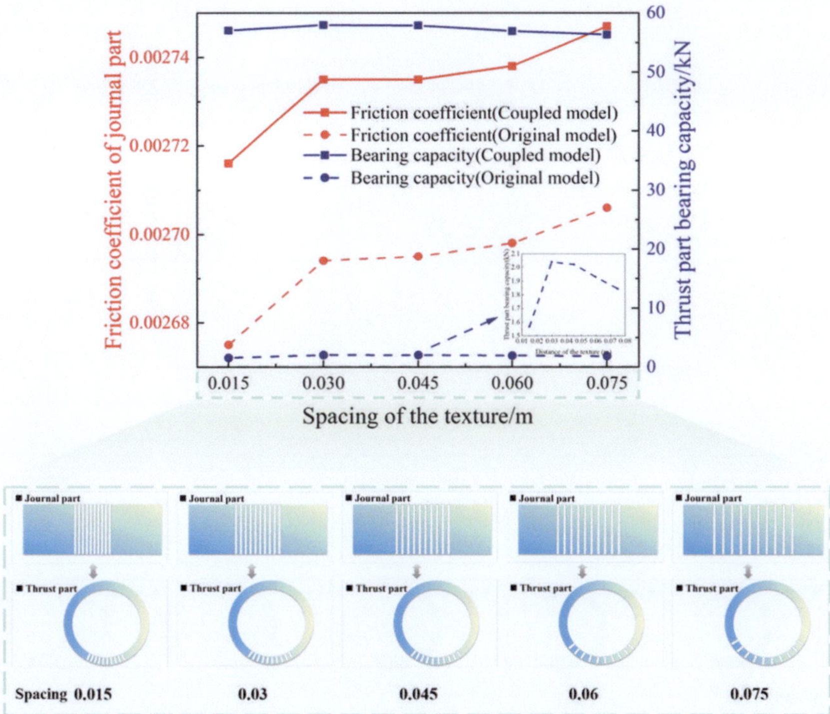

Fig. 4.9 Lubrication performance of the JTC bearing under different texture spacings

performance. The specific expression for R_A is as follows:

$$R_A = \max\left\{\frac{1}{a}\sum_{j=1}^{a} x_{ij}\right\} - \min\left\{\frac{1}{a}\sum_{j=1}^{a} x_{ij}\right\} \tag{4.15}$$

where a represents the number of repetitions for each level; x represents the test result, with subscript i means the number of levels and j means the number of repetitions.

The range analysis of the μ_J, as shown in the Fig. 4.12, reveals that T_D has the greatest impact on the μ_J. In addition, T_L, T_S and T_W affect the μ_J to a similar extent. Similarly, Fig. 4.12 shows the range analysis of the W_{th}, and T_L can be found to have the highest influence, followed by T_D, T_W and T_S. Therefore, when reducing μ_J is the research objective, T_D should be regarded as the main focus of the study; when increasing W_{th} is the research objective, T_L should be regarded as the main focus of the study. According to the calculated lubrication results, the variation in the μ_J is minimal, which could be attributed to the high load conditions. Therefore, the optimization of W_{th} should be taken as the leading role, and in the subsequent research, the T_L was selected as the optimization object of the PSO

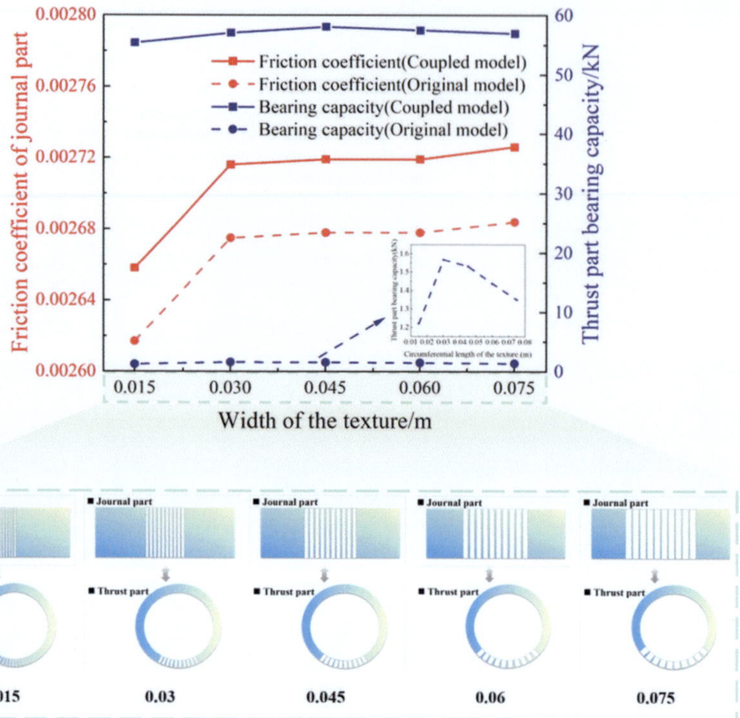

Fig. 4.10 Lubrication performance of the JTC bearing under different texture widths

Table 4.3 Parameters and levels of the orthogonal test

Parameter	Level 1	Level 2	Level 3	Level 4	Level 5
Depth (μm)	4	6	8	10	12
Length ratio	0.2	0.4	0.6	0.8	1
Spacing (m)	0.015	0.03	0.045	0.06	0.075
Width (m)	0.015	0.03	0.045	0.06	0.075

algorithm. According to the conclusion of the previous section, other parameters are designed respectively for subsequent PSO algorithm calculations: $T_D = 12$ μm, $T_S = 0.015$ m, $T_W = 0.03$ m.

Table 4.4 The orthogonal test schemes

Test number	Test schemes			
	Depth (μm)	Length ratio	Spacing (m)	Width (m)
T_1	4	0.2	0.015	0.015
T_2	4	0.4	0.045	0.06
T_3	4	0.6	0.075	0.03
T_4	4	0.8	0.03	0.075
T_5	4	1	0.06	0.045
T_6	6	0.2	0.075	0.06
T_7	6	0.4	0.03	0.03
T_8	6	0.6	0.06	0.075
T_9	6	0.8	0.015	0.045
T_{10}	6	1	0.045	0.015
T_{11}	8	0.2	0.06	0.03
T_{12}	8	0.4	0.015	0.075
T_{13}	8	0.6	0.045	0.045
T_{14}	8	0.8	0.075	0.015
T_{15}	8	1	0.03	0.06
T_{16}	10	0.2	0.045	0.075
T_{17}	10	0.4	0.075	0.045
T_{18}	10	0.6	0.03	0.015
T_{19}	10	0.8	0.06	0.06
T_{20}	10	1	0.015	0.03
T_{21}	12	0.2	0.03	0.045
T_{22}	12	0.4	0.06	0.015
T_{23}	12	0.6	0.015	0.06
T_{24}	12	0.8	0.045	0.03
T_{25}	12	1	0.075	0.075

4.3.3 *Texture Parameter Optimization Based on PSO Algorithm*

4.3.3.1 Research on Optimal Texture Optimization Under High Load

In contrast to conventional fixed-length texture designs, PSO algorithms enable individual design of the length for each texture. In addition, the PSO algorithm can take into account both the journal and thrust parts of the JTC bearing to achieve multi-objective optimization. To ensure population diversity and prevent excessive computation time, this work set the initial particle swarm size to 30 and the number of evolutionary generations to 60, following the reference (Zhang et al. 2021). In

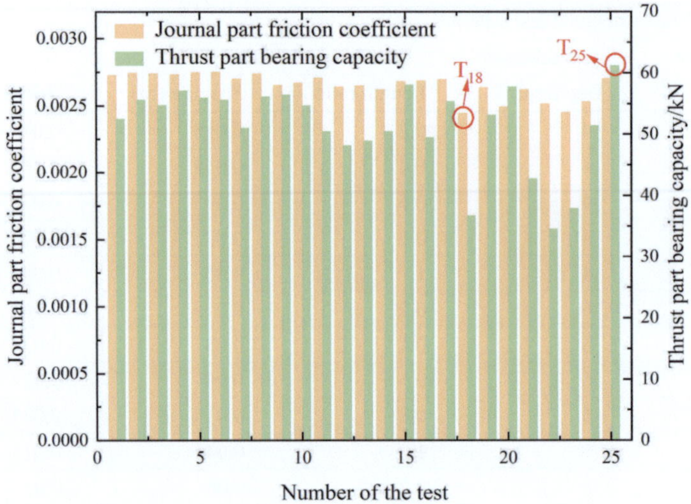

Fig. 4.11 The orthogonal test results: journal part friction coefficient and thrust part bearing capacity

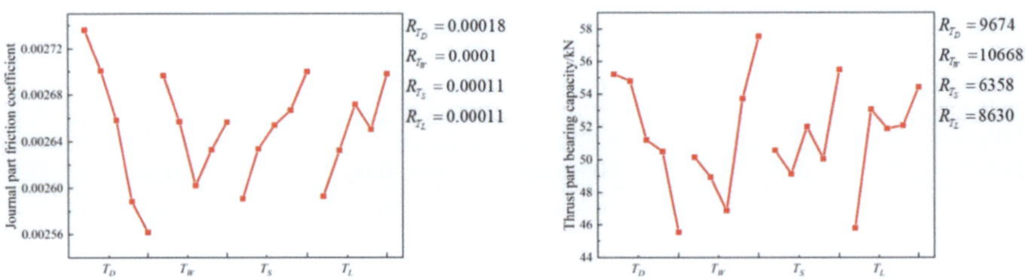

Fig. 4.12 The range analysis on journal part friction coefficient and thrust part bearing capacity

this study, the position vector x represents the texture length ratio (T_L), while the velocity vector v corresponds to the change in texture length ratio for each generation. The variable *id* denotes different textures on the bearing surface (1–10 for the texture in journal part, and 11–20 for the texture in thrust part). T corresponds to the evolution generation. $pbest_i{}^{\mathrm{T}}$ represents the partial optimal texture length ratio obtained when the i-th texture evolves to the T-th generation. $gbest^T$ represents the global optimal texture length ratio among all texture length groups when evolving to the T-th generation. For the journal part and the thrust part, the basis for judging the effect of texture length on lubrication performance is the friction coefficient and bearing capacity, respectively.

Firstly, texture design for JTC bearings under a high load of 600 kN was conducted using the PSO algorithm. Figure 4.13a illustrates the variation of the optimal thrust part bearing capacity (W_{Tbest}) solution with the evolution of generations. It can be observed that as the generation increases from 1 to 45, W_{Tbest} gradually increases,

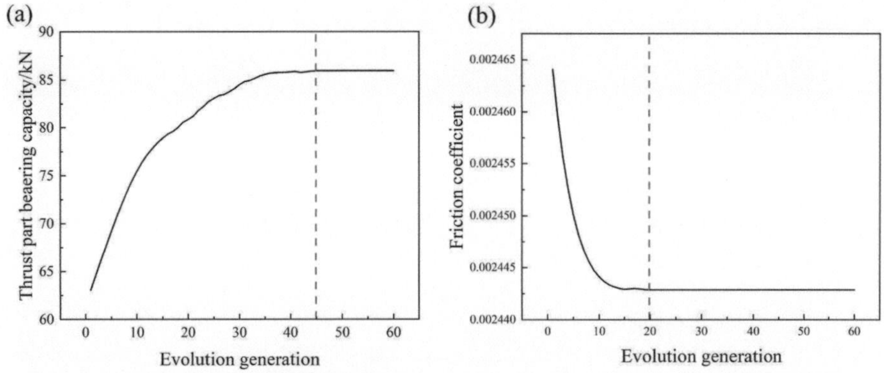

Fig. 4.13 Changing trend of the thrust part bearing capacity and journal part friction coefficient **a** thrust part bearing capacity; **b** journal part friction coefficient

indicating the continuous evolution of the optimal distribution of groove textures in thrust part. However, once the generation surpasses 45, W_{Tbest} stabilizes at a constant value, suggesting that the population evolution has been completed, and the solving process has reached a stable stage. Similarly, Fig. 4.13b shows the variation of the optimal journal part friction coefficient solution (μ_{best}) with the evolution of generations. It can be observed that as the generation rises from 1 to 20, μ_{best} gradually decreases, indicating the continuous evolution of the optimal distribution of groove textures in journal part. However, once the generation surpasses 45, W_{Tbest} stabilizes at a constant value, suggesting that the population evolution has been completed, and the solving process has reached a stable stage. Compared to the optimization of W_{th}, the optimal solution for μ_J can be found more quickly. This is because the influence of texture on the μ_J is relatively small under high eccentricity and the variation of μ_{best} in Fig. 4.13b can be found to be small.

The evolutionary process of the optimal distribution of groove textures, as shown in Fig. 4.14, begins with groove textures of random lengths and gradually evolves into the final optimal solution. It can be observed that, for the journal part, with the increase in generation, apart from the length of the texture, both the number and spacing of the textures undergo changes. Eventually, optimal result exhibits a non-uniform trapezoidal distribution of textures. However, judging from the numerical changes in μ_J, the effect of texture on μ_J under high load is not significant. As for the texture in the thrust part, the optimal result is relatively irregular while the changing trend of each texture is quite linear, which also shows that the PSO algorithm is a gradual optimization process.

Furthermore, the results of the film thickness and film pressure were analyzed. The results of the bearing with smooth surface were also calculated for comparative analysis. Since the journal part was subjected to a fixed load, the differences in the maximum pressures between the two surfaces were not significant, so the film thickness of the journal part was mainly analyzed. Figure 4.15 illustrates the distribution of film thickness for different surfaces. By comparing Fig. 4.15a and b, it can be

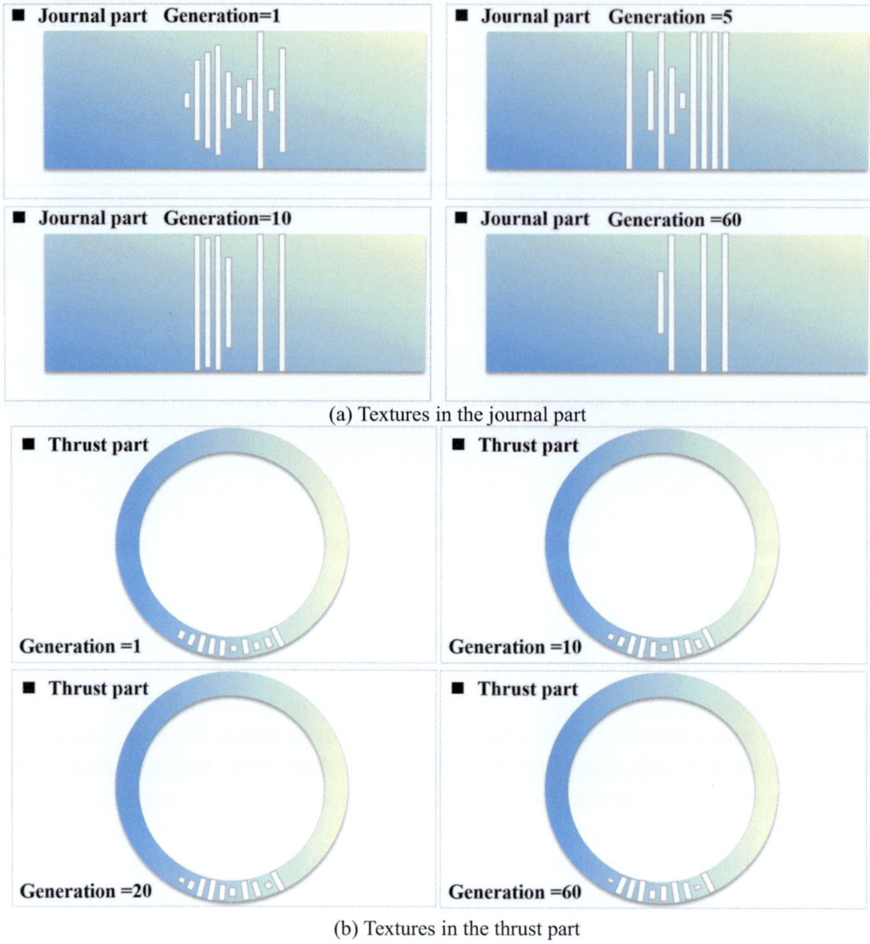

Fig. 4.14 The evolutionary process of the optimal distribution of groove textures

found that the bearing with optimized texture distribution has higher minimum film thickness compared to the one without texture, which reflects the positive effect of textures on increasing the film thickness. Figure 4.15b–e shows that the minimum film thickness gradually increases with the evolution of texture distribution, but the overall change is not significant. This is consistent with the change curve of the friction coefficient in Fig. 4.13, that is, the change in the film thickness of the journal part is small under this load, so the change in the friction coefficient is small.

For the thrust part, in order to explore the variation in its load-bearing capacity, the distributions of film pressure for different surface bearings were analyzed. A comparison of Fig. 4.16a and b reveals that the texture on the thrust part significantly increases the maximum film pressure from 32.9 to 38.6 MPa. Additionally, the pressure distribution shifts closer to the inner side at the 180° circumferential position

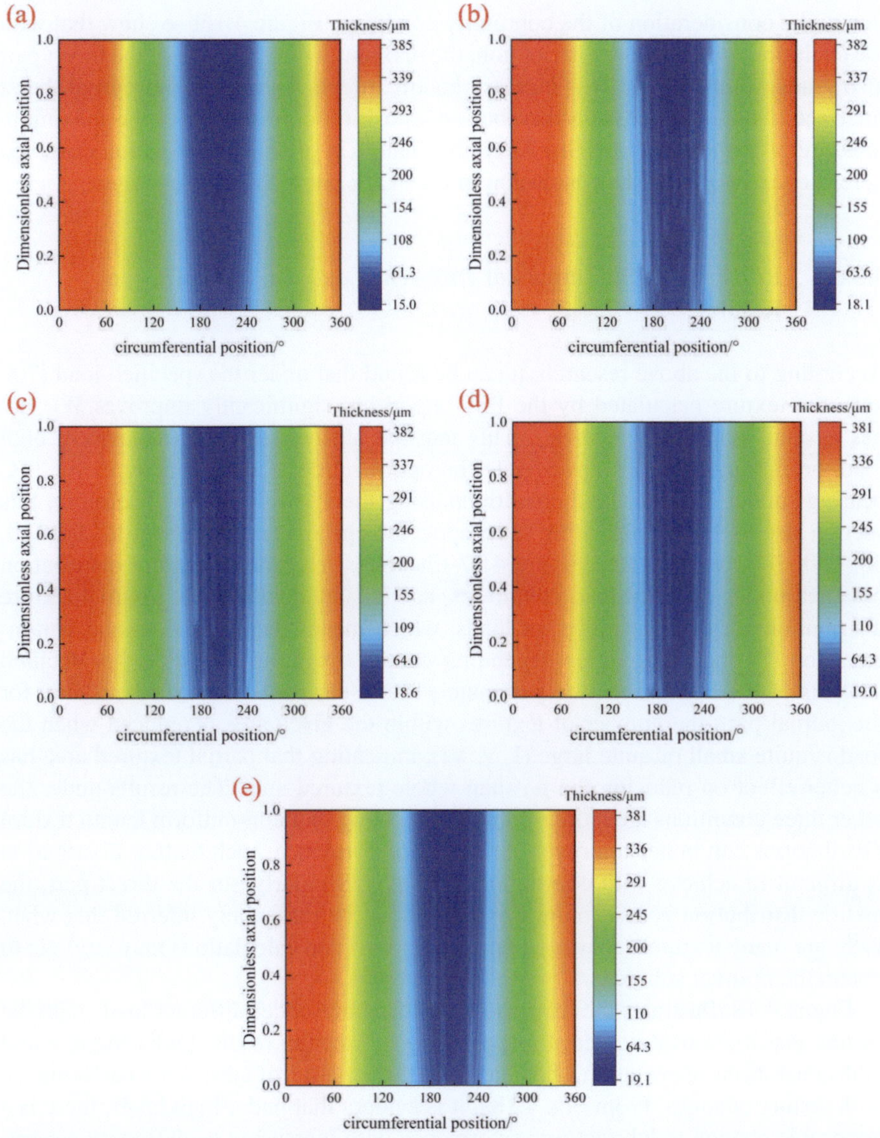

Fig. 4.15 Distribution of the journal part film thickness under different surface: **a** smooth surface; **b** textured surface (Generation = 1); **c** textured surface (Generation = 5); **d** textured surface (Generation = 10); **e** textured surface (Generation = 60)

due to the consideration of the continuity condition. Figure 4.16b–e show that with the evolution of the texture distribution, the maximum film pressure of the thrust part also gradually increases. This increase mainly affects the area at the journal-thrust interface, where the pressure distribution's position corresponds to the maximum pressure of the journal part. Similarly, the change in pressure distribution observed here is consistent with the conclusion in Fig. 4.13, but with larger changes.

4.3.3.2 Research on the Impact of Different Loads on Optimization Results

According to the above research, it can be found that under the specified load (700 kN), the texture calculated by the PSO algorithm significantly improves W_{th} but has a small improvement in μ_J. This may be attributed to the excessively high load, prompting a need to investigate the optimization of surface textures for JTC bearings under different load conditions using the PSO algorithm. Therefore, this section also used PSO algorithm to compute the optimal textures under loads of 10, 50, 100, 300, and 500 kN. Figure 4.17 illustrates the optimal texture distribution obtained under different load conditions, and it is evident that the optimal texture distribution varies under different loads, which means the optimal texture parameters for JTC bearings varies depending on the operating conditions. A detailed analysis of the optimal texture parameters under each load can be found that, for the journal part, the number of textures within the given area is reduced when the load is quite small or quite large (I, V, VI), indicating that partial textured area has a better effect on reducing the μ_J than whole textured area. The results under the other three conditions are more irregular, indicating that non-uniform length texture distributions can better improve lubrication performance, such texture distribution is difficult to achieve with conventional designs. Similarly, for the thrust part, the texture distribution is even more irregular. It can be reasonably inferred that when there are many textures, conventional assumptions and calculations may struggle to obtain the optimal solution.

Figure 4.18 illustrates the lubrication performance under different loads with the texture distributions designed. It presents the lubrication results under the first and 60th generations of evolution, characterizing the variation of lubrication performance with texture changes. From Fig. 4.18a, it is evident that under high loads, there is a noticeable change in lubrication performance with increasing evolutionary generations. Under a load of 700 kN, W_{th} increases from 63,051 N to 85,944 N, indicating that the texture has the greatest effect on W_{th} improvement. Conversely, under a load of 10 kN, W_{th} increases only from 8407 to 8805 N, showing a relatively smaller improvement in W_{th} due to the texture. Conversely, Fig. 4.18b indicates that under low loads, the change in the μ_J is more pronounced with increasing evolutionary generations. When the load is 10 kN, it decreases from 0.14954 to 0.14807, and when the load is 700 kN, it decreases from 0.0264 to 0.00263. This also shows that the small change in friction coefficient under high load is caused by the load, not that the PSO algorithm does not have an optimization effect.

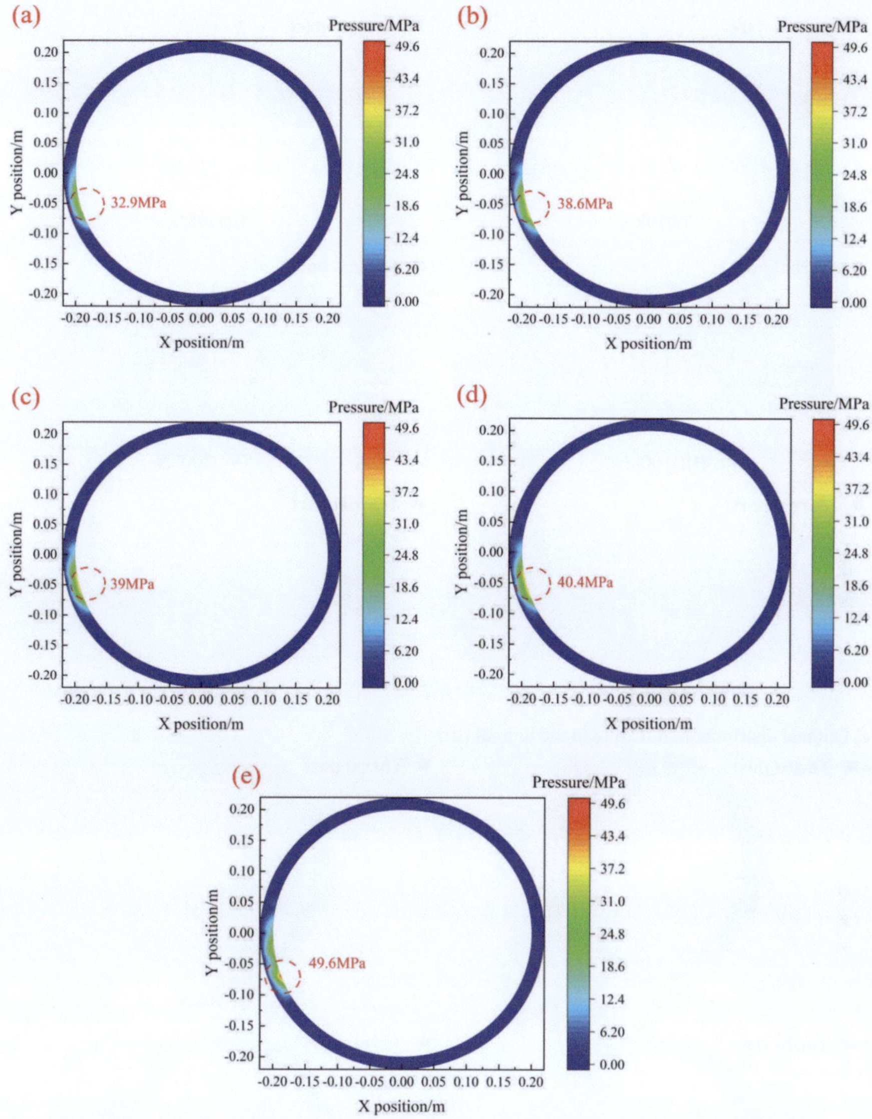

Fig. 4.16 Distribution of the thrust part film pressure under different surfaces: **a** smooth surface; **b** textured surface (Generation = 1); **c** textured surface (Generation = 10); **d** textured surface (Generation = 20); **e** textured surface (Generation = 60)

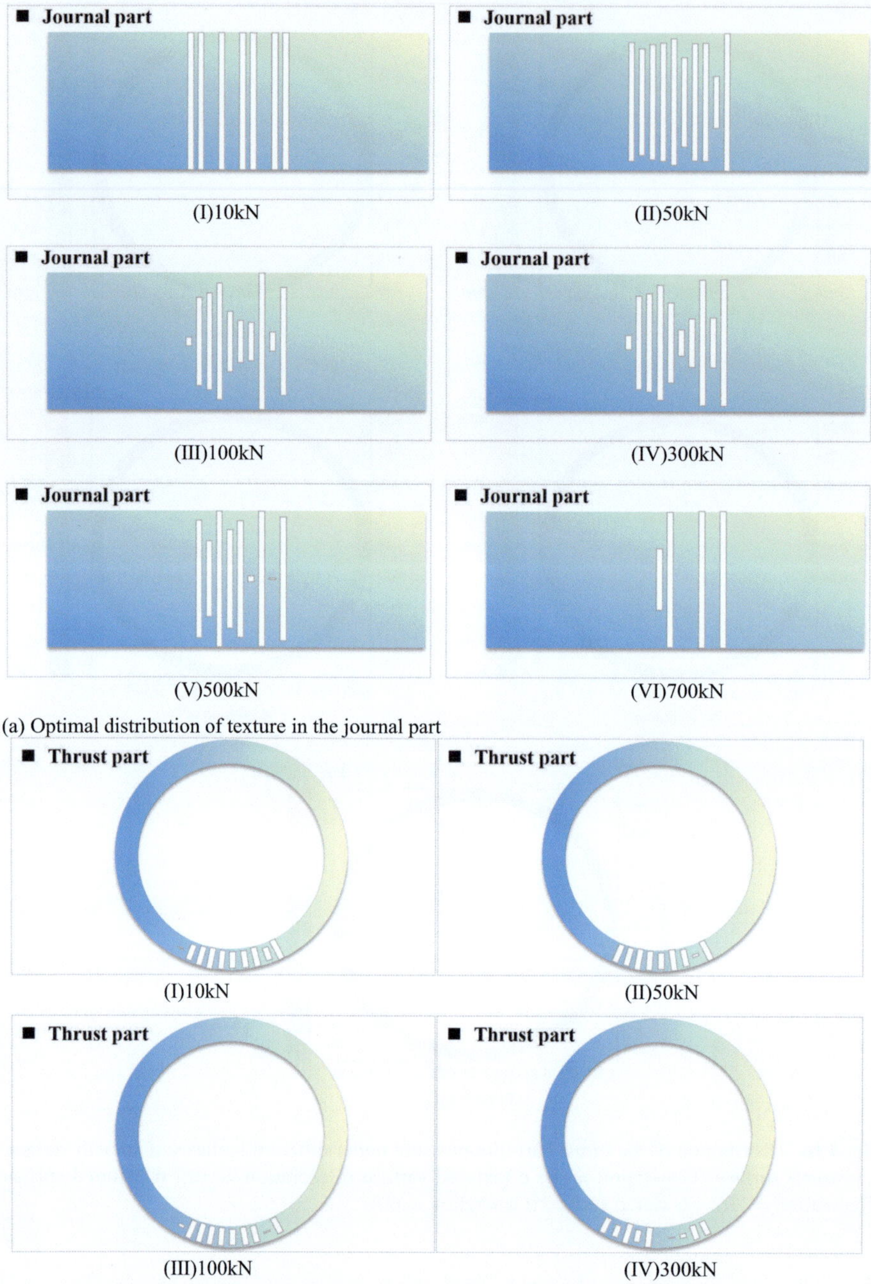

Fig. 4.17 Optimal distribution of texture under different load

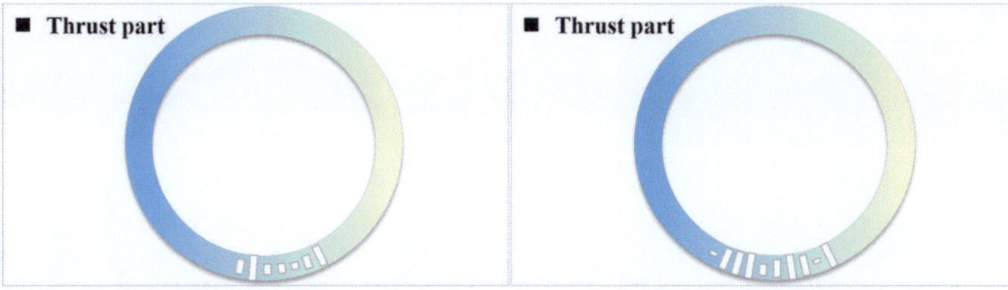

(V)500kN (VI)700kN

(b) Optimal distribution of texture in the thrust part

Fig. 4.17 (continued)

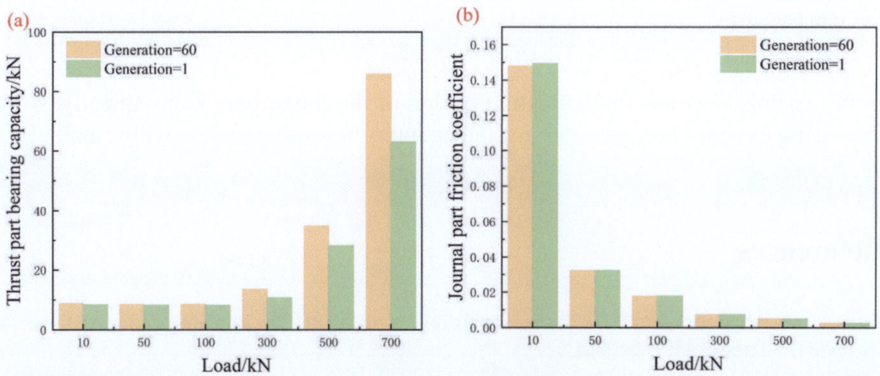

Fig. 4.18 Lubricant performance at the initial and final generations of evolution under different loads: **a** thrust part bearing capacity; **b** journal part friction coefficient

To further analyze the reasons, Fig. 4.19 compares the journal part film thickness distribution and thrust part pressure distribution under a 10 kN load with Fig. 4.16. Comparing Fig. 4.19a with Fig. 4.16a, it can be observed that under high loads, the minimum film thickness of the journal part is small, with a large gradient of thickness variation and concentrated pressure. Only the texture in the high-pressure area of the journal part affects lubrication, leading to a reduction in the effect of texture on the μ_J. Additionally, since the pressure in the journal part is significantly higher than that in the thrust part, W_{th} is greatly influenced by the pressure at the journal-thrust interface. Under the combined action of journal and thrust textures, W_{th} changes significantly. In contrast, comparing Fig. 4.19b with Fig. 4.16b, it can be seen that under low loads, the minimum film thickness of the journal part increases, with a smaller gradient of thickness variation and dispersed pressure. Almost all textures in the journal part are within the bearing area, leading to a significant improvement in μ_J due to texture. At this point, the pressure in the journal part is lower than that in the thrust part, and the distribution of thrust pressure matches the texture distribution. Therefore, the influence of journal part texture is relatively minor, and the increase

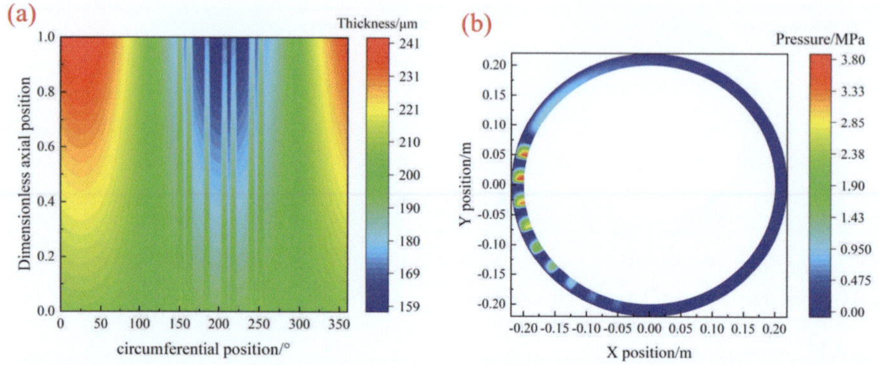

Fig. 4.19 Film thickness and film pressure under 10 kN load: **a** journal part film thickness; **b** thrust part film pressure

in W_{th} mainly depends on the texture design of the thrust part. Consequently, with increasing evolutionary generations, the changes in results are relatively small.

References

Allen Q, Raeymaekers B (2021) Surface texturing of prosthetic hip implant bearing surfaces: a review. J Tribol 143(4):040801

Chalkiopoulos M, Charitopoulos A, Fillon M et al (2020) Effects of thermal and mechanical deformations on textured thrust bearings optimally designed by a THD calculation method. Tribol Int 148:106303

Chan CW (2015) Modified particle swarm optimization algorithm for multi-objective optimization design of hybrid journal bearings. J Tribol 137(2):021101

Chen R, Zhao B, He T et al (2024) Study on coupling transient mixed lubrication and time-varying wear of main bearing in actual operation of low-speed diesel engine. Tribol Int 191:109159

Codrignani A, Savio D, Pastewka L et al (2020) Optimization of surface textures in hydrodynamic lubrication through the adjoint method. Tribol Int 148:106352

Galda L, Sep J, Olszewski A et al (2019) Experimental investigation into surface texture effect on journal bearings performance. Tribol Int 136:372–384

Gropper D, Wang L, Harvey TJ (2016) Hydrodynamic lubrication of textured surfaces: a review of modeling techniques and key findings. Tribol Int 94:509–529

Guo Z, Huang Q, Xie X et al (2021) Effects of spherical-platform texture parameters on the tribological performance of water-lubricated bearings. Wear 477:203863

Guo J, Ding B, Wang Y et al (2023). Co-optimization for hydrodynamic lubrication and leakage of V-shape textured bearings via linear weighting summation. Phys Scr 98(12):125218

Hamilton D, Walowit J, Allen C (1966) A theory of lubrication by microirregularities. J Fluid Eng. 88(1):177–185

Hu D, Guo Z, Xie X et al (2019) Effect of spherical-convex surface texture on tribological performance of water-lubricated bearing. Tribol Int 134:341–351

Ibatan T, Uddin MS, Chowdhury MAK (2015) Recent development on surface texturing in enhancing tribological performance of bearing sliders. Surf Coat Technol 272:102–120

Long R, Sun Y, Zhang Y et al (2024) Influence of micro-texture radial depth variations on the tribological and vibration characteristics of rolling bearings under starved lubrication. Tribol Int 194:109545

Mavrovouniotis M, Li C, Yang S (2017) A survey of swarm intelligence for dynamic optimization: algorithms and applications. Swarm Evol Comput 33:1–17

Pei S, Ma S, Xu H et al (2011) A multiscale method of modeling surface texture in hydrodynamic regime. Tribol Int 44(12):1810–1818

Pei S, Xu H, Yun M et al (2016) Effects of surface texture on the lubrication performance of the floating ring bearing. Tribol Int 102:143–153

Shi J, Zhao B, He T et al (2023) Tribology and dynamic characteristics of textured journal-thrust coupled bearing considering thermal and pressure coupled effects. Tribol Int 180:108292

Shi J, Zhao B, Tu L et al (2024) Transient lubrication analysis of journal-thrust coupled bearing considering time-varying loads and thermal-pressure coupled effect. Tribol Int 194:109502

Shi Y, Eberhart R (1998) A modified particle swarm optimizer. In: 1998 IEEE international conference on evolutionary computation proceedings. IEEE World Congress on Computational Intelligence, pp 69–73

Tala-Ighil N, Fillon M (2015) A numerical investigation of both thermal and texturing surface effects on the journal bearings static characteristics. Tribol Int 90:228–239

Wang W, Liu B, Zhang Y, Shao X, Allaire P (2018) Theoretical and experimental study on the static and dynamic characteristics of tilting-pad thrust bearing. Tribol Int 123:26–36

Wang W, He Y, Zhao J et al (2020) Optimization of groove texture profile to improve hydrodynamic lubrication performance: theory and experiments. Friction. 8(1):83–94

Wu C, Yang K, Ni J et al (2023) Investigations for vibration and friction torque behaviors of thrust ball bearing with self-driven textured guiding surface. Friction. 11(6):894–910

Xiang G, Han Y, Chen R et al (2020) A hydrodynamic lubrication model and comparative analysis for coupled microgroove journal-thrust bearings lubricated with water. P I Mech Eng J-J Eng. 234(11):1755–1770

Xiang G, Wang J, Zhou C et al (2021) A tribo-dynamic model of coupled journal-thrust water-lubricated bearings under propeller disturbance. Tribol Int 160:107008

Xie Z, Li J, Tian Y et al (2024) Theoretical and experimental study on influences of surface texture on lubrication performance of a novel bearing. Tribol Int 193:109351

Xu H, He T, Zhong N et al (2022) Transient thermomechanical analysis of micro cylindrical asperity sliding contact of SnSbCu alloy. Tribol Int 167:107362

Zhang H, Liu Y, Hua M et al (2018) An optimization research on the coverage of micro-textures arranged on bearing sliders. Tribol Int 128:231–239

Zhang X, Liu C, Zhao B (2021) An optimization research on groove textures of a journal bearing using particle swarm optimization algorithm. Mech Ind 22:1

Part II
Double-Layer Oil Film Floating Bushing in Diesel Engine Fuel Supply Mechanisms

Chapter 5
Overview of Double-Layer Oil Film Floating Bush for Diesel Engine Fuel Supply Mechanisms

Keywords Fuel pump · Floating bush · Double-layer oil film · Lubrication characteristics

5.1 Introduction

The high-pressure plunger pump, as the "power heart" of the diesel engine fuel system, is the core equipment supporting ships. Due to its high internal rated pressure, it has high requirements for component impact resistance, dynamics, and other performance, and the related technology has always been a bottleneck restricting the development of China's marine diesel engine fuel system. The high-pressure plunger pump of a marine diesel engine is the core component of its fuel injection system, and its core task is to deliver precisely metered fuel to each cylinder injector at extremely high pressure. It is usually driven by a camshaft and achieved through precision fitting plunger and roller components. When the cam pushes the plunger upward, the plunger compresses and generates enormous pressure. At a predetermined precise moment, a specific groove or control edge on the plunger aligns with the inlet/outlet hole on the sleeve (mechanical) or receives an electronic control signal (electronic control), allowing high-pressure fuel to overcome the spring force of the outlet valve and flow instantly through the high-pressure fuel pipe to the injector. This high pressure ensures that the fuel can be fully atomized when injected into the combustion chamber, efficiently mixed with compressed air, and completely burned, thereby providing strong and reliable propulsion power for large ships. In a high-pressure plunger pump, the reciprocating motion of the plunger determines the fuel injection timing and quantity. The cam mechanism is the core mechanical device that drives the plunger to generate high pressure (Kun 2023), and the roller floating liner pin structure is an important transmission structure of the cam mechanism, as shown in Fig. 5.1. During the working process, the camshaft drives the roller to rotate, while

B. Zhao et al., *Lubrication and Dynamic Characteristics Analysis Technology*,
https://doi.org/10.1007/978-981-95-2270-5_5

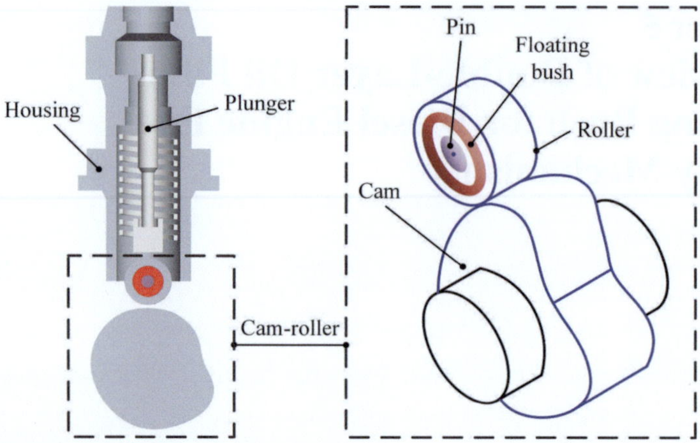

Fig. 5.1 Cam mechanism structure in fuel pump

the roller drives the floating bushing to rotate in the same direction. The roller pin is a fixed component, and a double-layer oil film can be formed inside and outside. The external load mainly comes from the plunger fuel pressure acting on the roller. At the same time, the rotational speed of the liner reduces the relative motion between the roller and the pin, reducing the heat and thermal deformation generated by friction, thereby reducing the locking rate (Zheng et al. 2020; Ma et al. 2021; Xie et al. 2023a, b, c).

5.2 Introduction of Steady-State Lubrication and Dynamic Characteristics of RFBP Structures

At present, scholars have carried out a large amount of research work on the lubrication performance of bearings with double-layer oil film structures (such as turbocharger float ring bearings, etc.), such as: analyzing the lubrication characteristics of float ring bearings with different structures (Soni et al. 2017; Pei et al. 2016; Yang et al. 2006; Wang et al. 2015); Fang (2016) studied the lubrication characteristics of floating ring bearings under oil-poor lubrication by using the CFD method and analyzed the influence of the degree of oil-poor on the lubrication characteristic parameters of the double-layer oil film. Li et al. (2017) studied the three-dimensional thermohydrodynamic lubrication (THD) characteristics and rotor vibration characteristics of floating ring bearings through simulation and experimental means; Xie et al. (2022) established a THD lubrication model for floating ring bearings and systematically analyzed the influence laws of vertical load and clearance ratio on lubrication characteristics. The above-mentioned research object is the floating ring bearing in the turbocharger. Most of them operate under high-speed and light-load conditions. Therefore, the floating ring is regarded as a rigid body, that is, the influence of elastic deformation is ignored. With the deepening of research, some scholars have also conducted studies on the elastic deformation in floating ring bearings: Li et al. (2017) established a lubrication analysis model for floating ring bearings considering thermal elastic deformation, explored the lubrication characteristics of floating ring bearings, and found that under high-speed and light-load conditions, compared with thermal deformation, the influence of elastic deformation of floating rings on lubrication characteristics can be ignored; When Qin et al. (2017) conducted the research using the numerical integration method, they considered the elastic deformation of the floating ring and believed that the elastic deformation began to have a certain impact on its lubrication characteristics only when the eccentricity was large.

5.3 Introduction of Transient Lubrication and Dynamic Characteristics of RFPB Structures

The roller-floating bushing-roller pin structure in the diesel engine fuel supply mechanism is a relatively typical double-layer oil film lubrication structure operating under low-speed and heavy-load conditions. The low-speed and heavy-load working conditions make the lubrication state rather poor, resulting in a mixed lubrication state (that is, the coexistence of micro-convex body contact and oil film lubrication). At the same time, elastic deformation and thermal effects are also more significant under heavy-load working conditions. However, at present, there are not many studies on the double-layer oil film lubricated bearing form under such low-speed and heavy-load working conditions. In Jigui (2010), a hydrodynamic lubrication model was established for the roller-floating bushing-roller pin structure in high-pressure oil pumps, and the load-bearing capacity of this structure was studied. Zhi (2022) established a fluid lubrication model of the double-layer oil film structure in the plunger pump based on the Fluent software, taking into account the influence of cavitation effect, but did not consider the effect of elastic deformation of the structure under low-speed and heavy-load conditions. In Limin et al. (2020), for the roller-floating bushing-roller pin structure of the diesel engine valve train, considering the effects of the mixed lubrication state and the elastic deformation of the bushing, a dynamic and lubrication coupling model was established based on commercial software and the structure was optimized accordingly. However, this study did not consider the effects of thermal effect and thermal deformation.

To sum up, the current research on the lubrication of double-layer oil film structures mostly focuses on the floating ring bearings under high-speed and light-load conditions, while there is relatively little analysis of the lubrication characteristics of the diesel engine roller-floating bushing-roller pin structure under low-speed and heavy-load conditions. Under high-speed and light-load conditions, the influence of structural elastic deformation can usually be ignored. Meanwhile, the lubrication state can be regarded as hydrodynamic lubrication, and the surface roughness characteristics can also be simplified for processing. However, the roller—floating bushing—roller pin and floating ring bearing have significant differences in structure and working conditions. Under low-speed and heavy-load working conditions, the influence of elastic deformation and roughness cannot be ignored. Therefore, in this paper, based on the average Reynolds equation of the double-layer oil film inside and outside the floating bushing, taking into account the elastic deformation of the floating bushing and the thermal deformation of the rollers, floating bushing and roller pins, and considering the influence of surface roughness under the mixed lubrication state, a mixed thermal elastohydrodynamic lubrication model of the roller—floating bushing—roller pin of the diesel engine fuel supply mechanism was constructed by means of designing simulation programs. The influence laws of different surface roughness of bushings, width-to-diameter ratio, gap ratio of inner and outer layers and bushing thickness on the lubrication characteristics were explored, which can provide theoretical support for the optimal design of the roller—floating bushing—roller pin structure.

References

Chao Q, Yang K, Kun S et al (2017) The influence of surface roughness on the lubrication static characteristics of floating ring bearings. Lubrication Sealing 42(5):66–72

Dequan Y, Yujie D (2006) Boundary element analysis of the stability of misaligned floating ring bearings. Lubrication Sealing 1:38–40

Fang L (2016) Analysis of lubrication characteristics of floating ring bearings under poor oil conditions. Hebei University of Technology, Tianjin

Jigui Y (2010) Research on the carrying capacity of sliding bearings in floating bushings of ultra-high pressure oil pumps. Shanghai Jiao Tong University, Shanghai

Kun D (2023) Research on the fuel supply performance and dynamics of high-pressure oil pumps. Harbin Engineering University

Li YJ, Liang F, Zhou Y et al (2017) Numerical and experimental investigation on thermohydrodynamic performance of turbocharger rotor-bearing system. Appl Therm Eng 121:27–38

Limin W, Mei L, Liting L et al (2020) Research on optimization of lubrication characteristics of diesel engine valve train roller floating ring bearing. Internal Combust Eng Eng 41(6):127–134

Ma J, Zhang H, Shi Z, Chu F, Gu F et al (2021) Modelling Acoustic Emissions induced by dynamic fluid-asperity shearing in hydrodynamic lubrication regime. Tribol Int 153:106590

Pei SY, Xu H, Yun M et al (2016) Effects of surface texture on the lubrication performance of the floating ring bearing. Tribol Int 102:143–153

Soni S, Vakharia DP (2017) Dynamic performance analysis of a noncircular cylindrical floating ring bearing. Proc Inst Mech Eng, Part J: J Eng Tribol 231(6):745–765

Wang LK, Bin GF, Li XJ et al (2015) Effects of floating ring bearing manufacturing tolerance clearances on the dynamic characteristics for turbocharger. Chinese J Mech Eng 28(3):530–540

Xie ZL, Zhu WD (2022) An investigation on the lubrication characteristics of floating ring bearing with consideration of multi-coupling factors. Mech Syst Signal Process 162:108086

Xie Z, Jiao J, Wrona S (2023a) The fluid-structure interaction lubrication performances of a novel bearing: experimental and numerical study. Tribol Int 179:108151

Xie Z, Jiao J, Yang K (2023b) A state-of-art review on the water-lubricated bearing. Tribol Int 180:108276

Xie Z, Jiao J, Yang K (2023c) Experimental and numerical study on the mixed lubrication performances of a new bearing. Tribol Int 182:108334

Zheng L, Zhu H, Zhu J, Deng Y (2020) Effects of oil film thickness and viscosity on the performance of misaligned journal bearings with couple stress lubricants. Tribol Int 146:106229
Zhi L (2022) Research on lubrication characteristics of floating ring bearings in high-pressure plunger pumps. Chongqing University of Technology, Chongqing

Chapter 6
Steady-State Thermal Elastic Hydrodynamic Lubrication and Dynamic Characteristics of Double-Layer Oil Film Floating Bush

Abstract Taking the roller—floating bush—roller pin in the fuel supply mechanism of Marine diesel engines as the research object, considering its elastic deformation and surface roughness characteristics under low-speed and heavy-load working conditions, a thermoelastic hydrodynamic lubrication analysis model of the double-layer oil film inside and outside the float bushing was established. The influence laws of the gap ratio of the inner and outer oil films, the width-to-diameter ratio of the bushing, the thickness of the bushing and its surface roughness on the lubrication characteristics such as the ring speed ratio, the maximum oil film pressure of the inner and outer layers and the bearing capacity were discussed. The results show that during the operation of the roller-float bushing—roller-pin structure, the thickness of the internal oil film is relatively small and the lubrication state is poor. Under low-speed and heavy-load working conditions, the elastic deformation of the bushing and the mixed lubrication effect cannot be ignored. The larger the width-to-diameter ratio and the smaller the gap ratio between the inner and outer layers, the smaller the maximum oil film pressure, the stronger the structural bearing capacity, but the higher the temperature rise. The mixed lubrication state only occurs within the inner lubricating oil film of the bushing. The greater the surface roughness of the bushing, the worse the lubrication effect.

Keywords Floating bush · Inner and outer double-layer oil film · Mixed thermoelastic fluid dynamic pressure lubrication analysis · Mixed lubrication

B. Zhao et al., *Lubrication and Dynamic Characteristics Analysis Technology*,
https://doi.org/10.1007/978-981-95-2270-5_6

6.1 Steady State Lubrication and Dynamic Characteristic Analysis Model Under the Coupling Effect of Double-Layer Oil Film

6.1.1 Analysis Model of Inner and Outer Oil Film Lubrication

The schematic diagram of the double-layer oil film of the RFBP structure is shown in Fig. 6.1. O_p is the center of the pin, O_1 and O_2 are the centers of the floating bush and roller, respectively. R_i and R_o are the inner and outer diameters of the floating bush, respectively. The average Reynolds Eq. (6.1) of the inner and outer oil films are used to solve the pressure (p_i, p_o) of the inner and outer oil films, as shown in Eqs. (6.1) and (6.2).

$$\frac{\partial}{R_i^2\partial\theta}\left(\phi_x\frac{\rho h_i^3}{\mu_i}\frac{\partial p_i}{\partial\theta}\right)+\frac{\partial}{\partial y}\left(\phi_y\frac{\rho h_i^3}{\mu_i}\frac{\partial p_i}{\partial y}\right)=6\omega_f\,\phi_c\frac{\partial(\rho h_i)}{\partial\theta}-6\omega_f\,\sigma_i\frac{\partial(\rho\phi_s)}{\partial\theta} \tag{6.1}$$

$$\begin{gathered}\frac{\partial}{R_o^2\partial\theta}\left(\phi_x\frac{\rho h_o^3}{\mu_o}\frac{\partial p_o}{\partial\theta}\right)+\frac{\partial}{\partial y}\left(\phi_y\frac{\rho h_o^3}{\mu_o}\frac{\partial p_o}{\partial y}\right)=6(\omega_f+\omega_2)\phi_c\frac{\partial(\rho h_o)}{\partial\theta}\\+6(\omega_f-\omega_2)\sigma_o\frac{\partial(\rho\phi_s)}{\partial\theta}\end{gathered} \tag{6.2}$$

where, ω_b and ω_2 are the speed of the floating bush and roller, respectively; ϕ_x and ϕ_y are pressure flow factors; ϕ_c is the contact factor; ϕ_s is the shear flow factor; σ_i and σ_o are the comprehensive roughness of the inner and outer surface, respectively; ρ is the density of the lubricating oil, and μ_i and μ_o are the viscosity of the lubricating oil considering the temperature viscosity effect of the inner and outer layers, respectively; h_i and h_o are the inner and outer film thickness, obtained by Eq. (6.3) and (6.4) respectively.

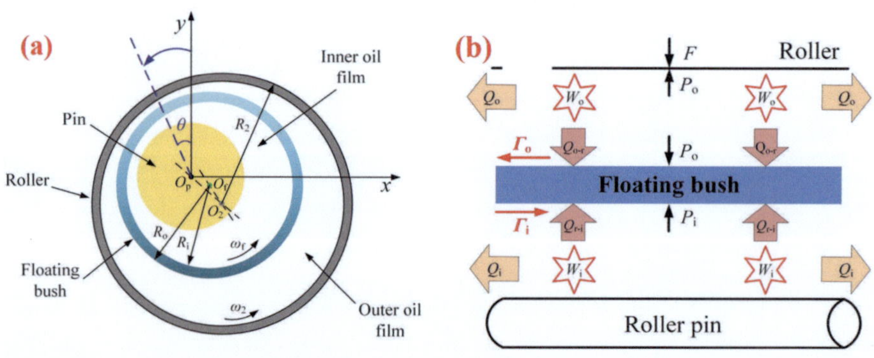

Fig. 6.1 Schematic diagram of double-layer oil film of the RFBP structure **a** Middle plane of the RFBP structure **b** Heat balance model

$$h_{\mathrm{i}} = (c_{\mathrm{i}} + \delta_{\mathrm{Ti}})(1 + \varepsilon_{\mathrm{i}}\cos(\theta - \varphi_{\mathrm{i}})) + \delta_{\mathrm{Pi}} \tag{6.3}$$

$$h_{\mathrm{o}} = (c_{\mathrm{o}} + \delta_{\mathrm{To}})(1 + \varepsilon_{\mathrm{o}}\cos(\theta - \varphi_{\mathrm{o}})) + \delta_{\mathrm{Po}} \tag{6.4}$$

where, ε_{i} and ε_{o} represent the eccentricity of the inner and outer oil film, respectively; δ_{Ti} and δ_{To} are the changes in gap caused by thermal deformation of the inner and outer layers, respectively, calculated by Eqs. (6.5) and (6.6).

$$\delta_{\mathrm{Ti}} = \left(\alpha_{\mathrm{f}} - \alpha_{\mathrm{p}}\right)R_{\mathrm{i}}\Delta T_{\mathrm{i}} \tag{6.5}$$

$$\delta_{\mathrm{To}} = (\alpha_{\mathrm{r}} - \alpha_{\mathrm{f}})R_{\mathrm{o}}\Delta T_{\mathrm{o}} \tag{6.6}$$

where, α_{P}, α_{B}, and α_{R} represent the thermal expansion coefficients of the pin, floating bush, and roller, respectively. ΔT_{i} and ΔT_{o} represent the temperature rise of the inner and outer oil films, respectively. ΔT_{b} is the temperature rise of the floating bush.

Under the slow-speed and heavy-load working conditions, the elastic deformation cannot be ignored in calculation. This work includes the elastic deformation of the floating bush.Elastic deformation of the floating bush (δ_{Pi}, δ_{Po}) can be obtained through Eqs. (6.7) and (6.8):

$$\delta_{\mathrm{P\,i}}(\theta, y) = \int_0^L \int_0^{2\pi} KI(\theta, y) p_i(\theta, y) \mathrm{d}\theta \mathrm{d}y \tag{6.7}$$

$$\delta_{\mathrm{Po}}(\theta, y) = \int_0^L \int_0^{2\pi} KO(\theta, y) p_o(\theta, y) \mathrm{d}\theta \mathrm{d}y \tag{6.8}$$

where $\delta_{\mathrm{pi}}(\theta,y)$ and $\delta_{\mathrm{po}}(\theta,y)$ are the elastic deformations of the inner and outer layers of the floating bushing at the node (θ, y) respectively. $p_{\mathrm{i}}(\theta, y)$, $p_{\mathrm{o}}(\theta, y)$ are the oil film pressure at node (θ, y) for the inner and outer layers of the floating bushing respectively.

When the floating bush works, it is mainly affected by the frictional force of the inner and outer oil films. The calculation formula for the viscous friction torque of the inner and outer layers is shown in Eqs. (6.9) and (6.10). The end leakage flow caused by the axial pressure gradient of the inner and outer oil films can be calculated by Eqs. (6.11) and (6.12), and the frictional power consumption generated by the shear force of the inner and outer oil films can be calculated by Eqs. (6.13) and (6.14).

$$\Gamma_{\mathrm{i}} = \int_0^L \int_0^{2\pi} \left\{ -\frac{h_i}{2} \frac{\partial p_i}{R_i \partial \theta} - \frac{\mu_i}{h_i} \omega_r R_i \right\} R_i^2 \mathrm{d}\theta \mathrm{d}y \tag{6.9}$$

$$\Gamma_o = \int_0^L \int_0^{2\pi} \left\{ \frac{h_o}{2} \frac{\partial p_o}{R_o \partial \theta} + \frac{\mu_o}{h_o} (\omega_r - \omega_b) R_o \right\} R_o^2 \mathrm{d}\theta \mathrm{d}y \tag{6.10}$$

$$Q_i = 2 \int_0^{2\pi R_i} \left[-\frac{h_i^3}{12\mu_i} \frac{\partial p_i}{\partial y} \right]_{y=L} \mathrm{d}x \tag{6.11}$$

$$Q_o = 2 \int_0^{2\pi R_o} \left[-\frac{h_o^3}{12\mu_o} \frac{\partial p_o}{\partial y} \right]_{y=L} \mathrm{d}x \tag{6.12}$$

$$W_i = \int_0^{L_i} \int_0^{2\pi R_i} \left\{ \frac{h_i^3}{2} \left[\left(\frac{\partial p_i}{\partial x} \right)^2 + \left(\frac{\partial p_i}{\partial y} \right)^2 \right] + \frac{\mu_i}{h_i} \omega_r^2 R_i^2 \right\} \mathrm{d}x\mathrm{d}y \tag{6.13}$$

$$W_o = \int_0^{L_o} \int_0^{2\pi R_o} \left\{ \frac{h_o^3}{2} \left[\left(\frac{\partial p_o}{\partial x} \right)^2 + \left(\frac{\partial p_o}{\partial y} \right)^2 \right] + \frac{\mu_o}{h_o} (\omega_b - \omega_r)^2 R_o^2 \right\} \mathrm{d}x\mathrm{d}y \tag{6.14}$$

In the mixed lubrication state caused by low-speed heavy load conditions, there exists micro convex contact force. Assuming that the probability density function of the height distribution of micro convex bodies on the contact surface is a normal Gaussian distribution expected to be equal to zero, and the surface morphology is isotropic, the contact pressure pc per unit area of micro convex bodies can be calculated by Eq. (6.15) (Pei et al. 2016).

$$p_c = \left(\frac{16\sqrt{2}}{15} \right) \pi (\eta \beta \sigma)^2 E' \sqrt{\frac{\sigma}{\beta}} F_{5/2}(H) \tag{6.15}$$

where E' is the composite elastic modulus; σ is the comprehensive surface roughness; η is the peak element density; β is the peak element curvature radius; H is the film thickness ratio, which is the ratio of oil film thickness to overall surface roughness.

When reaching the state of balance under stable external load conditions, the sum of the outer oil film reaction force and the outer contact pressure is equal to the sum of the inner oil film reaction force and inner the contact pressure, while also balancing with the external load. In addition, the friction torque of the inner oil film is balanced with that of the outer oil film. The balanceable relationship is shown in Eq. (6.16) and (6.17):

$$P_{\mathrm{apsi}} + P_{\mathrm{i}} = P_{\mathrm{apso}} + P_{\mathrm{o}} = F \tag{6.16}$$

$$\Gamma_{\mathrm{i}} = \Gamma_{\mathrm{o}} \tag{6.17}$$

where P_{aspi} and P_{aspo} respectively represent the bearing capacity of the inner and outer micro convex bodies; P_{i} and P_{o} are the internal and external oil film reaction forces, respectively.

In addition, the heat transfer between the inner and outer oil films of the floating liner is equal, and the inner and outer oil films should meet the thermal equilibrium conditions respectively, that is, satisfy the following relationship:

$$c_{\mathrm{p}}\rho Q_{\mathrm{i}}\Delta T_{\mathrm{i}} + A_{\mathrm{Ri}}H_{\mathrm{Ri}}(T_{\mathrm{i}} - T_{\mathrm{R}}) = W_{\mathrm{i}} \tag{6.18}$$

$$c_{\mathrm{p}}\rho Q_{\mathrm{o}}\Delta T_{\mathrm{o}} + A_{\mathrm{Ro}}H_{\mathrm{Ro}}(T_{\mathrm{o}} - T_{\mathrm{R}}) = W_{\mathrm{o}} \tag{6.19}$$

$$A_{\mathrm{Ri}}H_{\mathrm{Ri}}(T_{\mathrm{R}} - T_{\mathrm{i}}) = A_{\mathrm{Ro}}H_{\mathrm{Ro}}(T_{\mathrm{o}} - T_{\mathrm{R}}) \tag{6.20}$$

where A_{Ri} and A_{Ro} are the heat transfer areas on the inner and outer sides of the floating bush, respectively, and W_{i} and W_{o} are the friction losses on the inner and outer layers, respectively; Q_{i} and Q_{o} respectively refer to the leakage flow at the inner and outer layers; T_{B} is the temperature of the floating bush; c_{p} is the specific heat of lubrication oil; H_{Ri} and H_{Ro} are the convective heat transfer coefficients of the inner and outer layers of the floating bush, respectively. The oil film temperature T can be calculated by Eq. (6.21); The convective heat transfer coefficient HR can be calculated by Eq. (6.22).

$$H_R = 3\mathrm{Pr}^{1/3}(\kappa/c) \tag{6.21}$$

$$T = T_{\mathrm{s}} + k\Delta T \tag{6.22}$$

where T_{s} is the inlet oil temperature. Under low-speed and heavy-load conditions, conduction heat dissipation needs to be considered, with a k value of 0.8; κ is the heat conductivity coefficient of lubricating oil; c is the radius clearance. Pr is the Planck number, which can be calculated by Eq. (6.23).

$$\mathrm{Pr} = c_{\mathrm{p}}\mu/\kappa \tag{6.23}$$

where μ is the viscosity of the lubricating oil.

6.1.2 Dynamic Characteristic Calculation Model

Figure 6.2 is a schematic diagram of the dynamic characteristic model, and the principle of the micro disturbance method is shown in Fig. 6.2a. Where, the inner oil film is taken as an example. The center of the floating bush (O_{f}) reaches O_{x1} and O_{x2} after a disturbance displacement of $-\Delta x$ and $+\ \Delta x$ in the x-direction. The

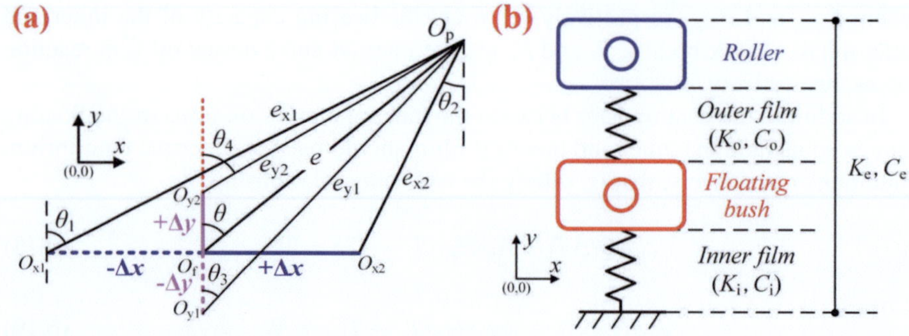

Fig. 6.2 Dynamic characteristic model of the RFBP structure. **a** Schematic diagram of micro disturbance method **b** Equivalent dynamic characteristics

eccentricities after disturbance are e_{x1} and e_{x2}. θ_1 and θ_2 are the starting and ending angles of the oil film, which refer to the direction angles after disturbance is generated. After disturbance displacement of $-\Delta y$ and $+ \Delta y$ in the y-direction, the center of the floating bush reaches O_{y1} and O_{y2}. The eccentricities after disturbance are e_{y1} and e_{y2}. θ_3 and θ_4 are the starting and ending angles of the oil film. θ' and θ'' are used to uniformly represent the oil film position angle before and after disturbance as follows, and the stiffness coefficient of the inner oil film is calculated by Eq. (6.24):

$$\begin{bmatrix} K_{xx} \; K_{yx} \\ K_{xy} \; K_{yy} \end{bmatrix}_i = \int_0^L \int_{\theta'}^{\theta''} - \begin{bmatrix} p_x \\ p_y \end{bmatrix}_i [\cos\theta \;\; \sin\theta] \mathrm{d}x\mathrm{d}y \tag{6.24}$$

where p_x and p_y the ratios of oil film pressure to Δx and Δy, respectively, and L is the width of the floating bush. The stiffness coefficient of the outer oil film is calculated by Eq. (6.25):

$$\begin{bmatrix} K_{xx} \; K_{yx} \\ K_{xy} \; K_{yy} \end{bmatrix}_o = \int_0^L \int_{\theta'}^{\theta''} - \begin{bmatrix} p_x \\ p_y \end{bmatrix}_o [\cos\theta \;\; \sin\theta] \mathrm{d}x\mathrm{d}y \tag{6.25}$$

The disturbance velocity method is used to calculate the damping coefficients of the oil films, and, the damping coefficients of the inner/outer oil film are calculated by Eqs. (6.26) and (6.27) respectively.

$$\begin{bmatrix} C_{xx} \; C_{yx} \\ C_{xy} \; C_{yy} \end{bmatrix}_i = \int_0^L \int_{\theta'}^{\theta''} - \begin{bmatrix} p_u \\ p_v \end{bmatrix}_i [\cos\theta \;\; \sin\theta] \mathrm{d}x\mathrm{d}y \tag{6.26}$$

$$\begin{bmatrix} C_{xx} & C_{yx} \\ C_{xy} & C_{yy} \end{bmatrix}_o = \int_0^L \int_{\theta'}^{\theta''} -\begin{bmatrix} p_u \\ p_v \end{bmatrix}_o [\cos\theta \ \sin\theta] dxdy \quad (6.27)$$

where p_u and p_v are the ratios of oil film pressure to x and y direction disturbance velocities, respectively. In addition, the main stiffness coefficients are expressed as K_{xx} and K_{yy}, while K_{xy} and K_{yx} are the cross stiffness coefficients. The main damping coefficients are expressed as C_{xx} and C_{yy}, while C_{xy} and C_{yx} are the cross damping coefficients.

The RFBP structure can be treated as two fluid film bearings in series (as can be seen Fig. 6.2b). The stiffness coefficients of the inner/outer films are expressed by K_i and K_o in Fig. 6.2b, respectively, and the damping coefficients of the inner/outer films are expressed by C_i and C_o, respectively. In addition, the equivalent stiffness/damping coefficients of the RFBP structure are expressed by K_e and C_e, respectively. The main/cross stiffness coefficients and the main/cross damping coefficients can be obtained as follows:

$$\begin{Bmatrix} \frac{1}{K_{xx}} & \frac{1}{K_{xy}} \\ \frac{1}{K_{yx}} & \frac{1}{K_{yy}} \end{Bmatrix}_e = \begin{Bmatrix} \frac{1}{K_{xx}} & \frac{1}{K_{xy}} \\ \frac{1}{K_{yx}} & \frac{1}{K_{yy}} \end{Bmatrix}_i + \begin{Bmatrix} \frac{1}{K_{xx}} & \frac{1}{K_{xy}} \\ \frac{1}{K_{yx}} & \frac{1}{K_{yy}} \end{Bmatrix}_o \quad (6.28)$$

$$\begin{Bmatrix} \frac{1}{C_{xx}} & \frac{1}{C_{xy}} \\ \frac{1}{C_{yx}} & \frac{1}{C_{yy}} \end{Bmatrix}_e = \begin{Bmatrix} \frac{1}{C_{xx}} & \frac{1}{C_{xy}} \\ \frac{1}{C_{yx}} & \frac{1}{C_{yy}} \end{Bmatrix}_i + \begin{Bmatrix} \frac{1}{C_{xx}} & \frac{1}{C_{xy}} \\ \frac{1}{C_{yx}} & \frac{1}{C_{yy}} \end{Bmatrix}_o \quad (6.29)$$

The relative units of dimensionless stiffness coefficients of the inner/outer films are taken as $\mu_i\omega_f L/\psi_i^2$, $\mu_o(\omega_2\text{-}\omega_f)L/\psi_o^2$. The relative units of dimensionless damping coefficients of the inner/outer films are taken as $\mu_i L/\psi_i^2$, $\mu_o L/\psi_o^2$. The inner film dimensionless stiffness coefficients are expressed as $\overline{K}_{i,xx}$, $\overline{K}_{i,xy}$, $\overline{K}_{i,yx}$, $\overline{K}_{i,yy}$, and $\overline{K}_{o,xx}, \overline{K}_{o,xy}, \overline{K}_{o,yx}, \overline{K}_{o,yy}$ are the outer film dimensionless stiffness coefficients. The inner film dimensionless damping coefficients are expressed as $\overline{C}_{i,xx}, \overline{C}_{i,xy}, \overline{C}_{i,yx}, \overline{C}_{i,yy}$, and $\overline{C}_{o,xx}, \overline{C}_{o,xy}, \overline{C}_{o,yx}, \overline{C}_{o,yy}$ are the outer film dimensionless damping coefficients. In addition, the equivalent stiffness coefficients of the RFBP structure are $\overline{K}_{e,xx}, \overline{K}_{e,xy}, \overline{K}_{e,yx}, \overline{K}_{e,yy}$, and the equivalent damping is $\overline{C}_{e,xx}$, $\overline{C}_{e,xy}, \overline{C}_{e,yx}, \overline{C}_{e,yy}$.

6.2 Solution Procedure

Figure 6.3 shows the flow chart of thermal elastohydrodynamic lubrication analysis for the mixture of the RFBP structure. Firstly, the outer film eccentricity, displacement angle, and temperature rise are determined, and then, the inner film eccentricity, displacement angle, and temperature rise are determined. Finally, the speed and temperature of the floating bush are determined by the balance of the friction torque and heat balance. The finite difference method is used for numerical solution, and

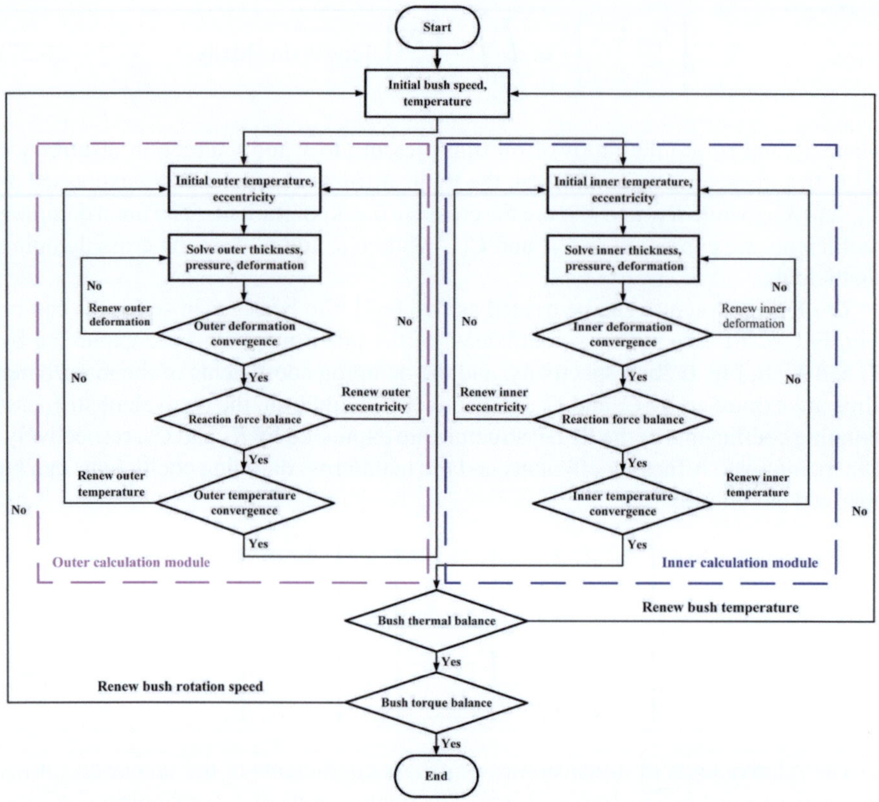

Fig. 6.3 Flow chart of steady-state lubrication numerical analysis of RFBP structure

the convergence accuracy of pressure solution is 10^{-5}. The number of grids is 120 $\times$ 16 (circumference $\times$ axial), and the super relaxation iteration method is used to accelerate convergence.

6.3 Model Validation

To verify the accuracy of the model, the calculation results of the floating ring bearing in reference (Yang Dequan et al. 2006; Wang et al. 2015; Liu Fang 2016) were compared and validated. The comparison results are shown in Fig. 6.4. Figure 6.4a, b show the comparison between the ring speed ratio (the ratio of floating ring speed to journal speed) and the outlet oil temperature calculated by the model in this paper and the results in the literature. It can be found that the deviation between the ring speed ratio and outlet oil temperature calculated by the model in this paper and the results in the literature do not exceed 5%, which verifies the accuracy of the model in this paper.

In order to verify the accuracy of the calculation model of dynamic characteristic parameters, the single-layer oil film simulation model, which is the degradation model of this work, was compared with the calculation results in the literature (Li et al. 2017). The comparison results are shown in Fig. 6.5. It can be observed that the trend of the two results is basically consistent, that is, as the eccentricity increases, both the oil film stiffness and oil film damping increase. The deviation between the two is within 8%, indicating the accuracy of the oil film dynamic characteristic model in this work.

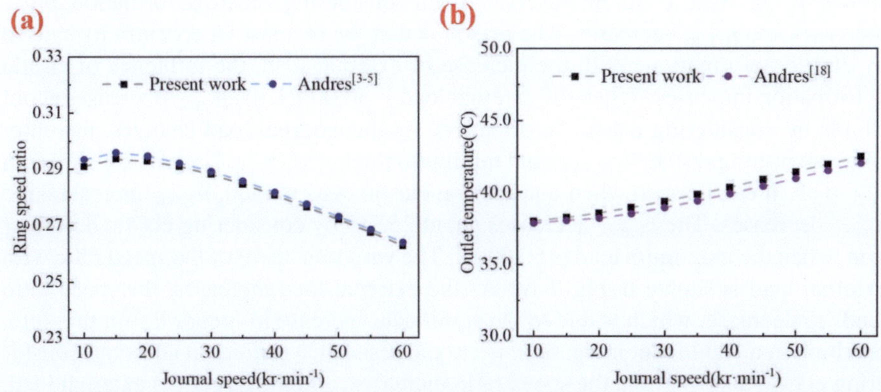

Fig. 6.4 The verification of steady-state model **a** Ring speed ratio; **b** Outer temperature

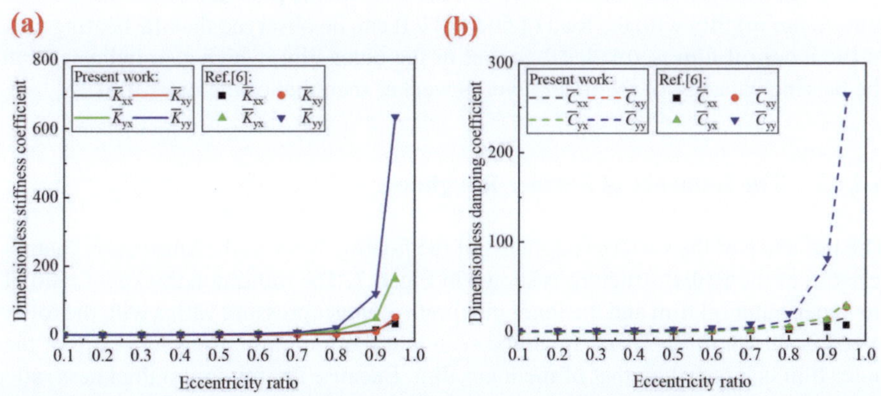

Fig. 6.5 Verification results of the dynamic characteristic parameters. **a** Variation of dimensionless stiffness coefficient; **b** Variation of dimensionless damping coefficient

6.4 Results Analysis

6.4.1 *Lubrication Characteristics Analysis of the RFBP Structure*

6.4.1.1 The Influence of Elastic Deformation

The influence of elastic deformation of floating bush on the lubrication characteristics of RFBP structure is shown in Fig. 6.6. As the external load changes, the inner film maximum pressure ($p_{i,max}$) and minimum thickness ($h_{i,min}$) curves are shown in Fig. 6.6a. It can be observed when considering elastic deformation, $p_{i,max}$ increases and $h_{i,min}$ decreases. The reason is that the increase in eccentricity caused by elastic deformation. With the increase of external load, the influence of elastic deformation increases. When the external load is 50,000N, the $p_{i,max}$ decreases about 51.1% by considering elastic deformation. As the external load changes, the outer film maximum pressure ($p_{o,max}$) and minimum thickness ($h_{i,min}$) curves are shown in Fig. 6.6b. It can be seen when considering elastic deformation, $p_{o,max}$ increases and $h_{o,min}$ decreases. The $p_{o,max}$ decreases about 25.1% by considering elastic deformation, when the maximum load is 50,000N. The variation curve of the speed ratio with external load is shown in Fig. 6.6c. As the external load increases, the speed ratio tends to decrease, which is due to the significant increase in inner oil film pressure, resulting in a significant increase in frictional resistance torque. In addition, considering elastic deformation, the speed ratio increases, and the larger the external load, the more significant the increase in speed ratio. The reason is that the decrease in oil film pressure in the inner layer is more significant than that in the outer layer when elastic deformation is considered. Figure 6.6d shows the pressure distribution of the inner/outer oil film with the load of 50,000N. It can be observed that the bearing area of the inner oil film is smaller than that of the outer film, which also indicates that the bearing capacity of the inner layer is weaker than that of the outer film.

6.4.1.2 The Influence of Surface Roughness

The influence of the surface roughness of the floating bush on the lubrication characteristics of the RFBP structure is shown in Fig. 6.7. The minimum thickness ratio of the inner/outer oil film and the inner maximum contact pressure varies with the roller speed is shown in Fig. 6.7a. It can be seen that the minimum thickness ratio of the outer film is larger than that of the inner film. Because the minimum thickness ratio of the inner film is less than 4 times the comprehensive roughness, the inner oil film is in a mixed lubrication state, while the outer layer is in a full film lubrication state. The maximum micro convex contact pressure in the inner layer increases with the increase of roller speed. When the roller speed is 1,000 r/min, the maximum micro convex contact pressure is about 5.11 MPa. The variation of the minimum thickness of the inner/outer film with the roller speed is shown in Fig. 6.7b, c. The minimum

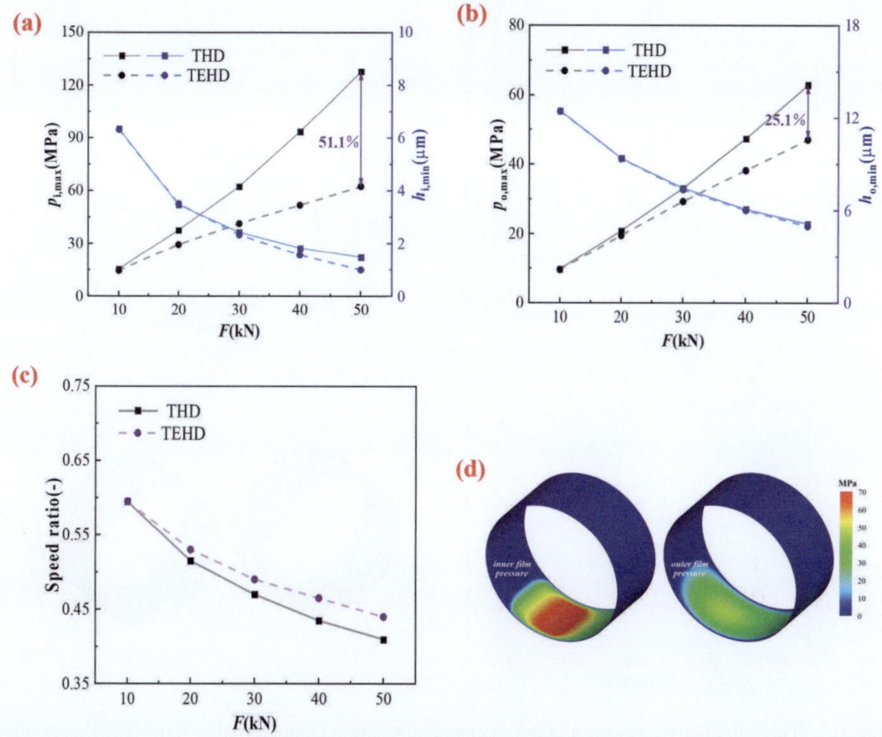

Fig. 6.6 The influence of elastic deformation of floating bush on the lubrication characteristics of the RFBP structure **a** $p_{i,max}$ and $h_{i,min}$ **b** $p_{o,max}$ and $h_{o,min}$ **c** Speed ratio **d** Oil film pressure distribution

oil film thickness of the inner layer increases by considering surface roughness, and the smaller the speed, the more obvious the increase. However, considering the surface roughness has no effect on the minimum oil film thickness of the outer layer. Figure 6.7d shows the pressure distribution of the inner/outer oil film with the roller speed of 1000 r/min. It can be observed that due to the influence of elastic deformation, the contact pressure of the inner micro convex body is mainly distributed at the axial edge position, while the outer layer is in a fully lubricated state, so there is no micro convex body contact force in the outer layer.

6.4.1.3 The Influence of Inner Eccentricity on Lubrication Characteristics

The variation curve of the lubrication characteristics of the RFBP structure which change with the eccentricity of the inner oil film, is shown in Fig. 6.8. The variation curve of the minimum thickness of the inner/outer film with the inner film eccentricity is shown in Fig. 6.8a. And the variation curve of the outer film eccentricity with the

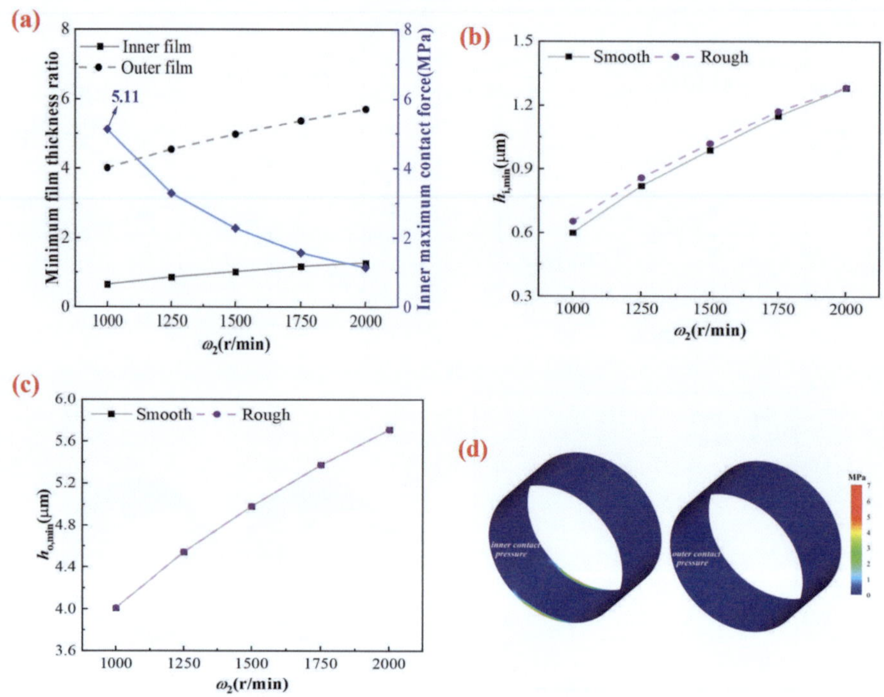

Fig. 6.7 The influence of surface roughness on the lubrication characteristics of the RFBP structure **a** Minimum film thickness ratio and inner maximum contact force **b** Minimum thickness of inner film **c** Minimum thickness of outer film **d** Contact pressure distribution

inner film eccentricity is also shown in Fig. 6.8a. It can be seen that as the eccentricity of the inner film increases, the minimum thickness of both the inner and outer film decreases, and the decrease in film minimum thickness of the inner film is more obvious. It can be seen that when stable state is reached, the inner film eccentricity is greater than the outer film eccentricity, and as the inner film eccentricity increases, the outer film eccentricity also increases, but the increase is smaller than that of the inner film. When the inner film eccentricity reaches 0.9, the outer film eccentricity only reaches 0.39, which is due to the weaker load-bearing capacity of the inner layer structure. The variation curve of the maximum pressure of the inner/outer film with the eccentricity of the inner film is shown in Fig. 6.8b. It can be seen that as the eccentricity of the inner film increases, the maximum pressure of both the inner and outer film increases, and the inner film maximum pressure increases more obviously. The curve of the variation of the speed ratio with the eccentricity of the inner film is shown in Fig. 6.8c. As the eccentricity of the inner film increases, the speed ratio shows a downward trend, because of the eccentricity of the inner film increases. The maximum pressure of the inner film increases more significantly than that of the outer film, resulting in a more significant increase in the inner layer resistance torque.

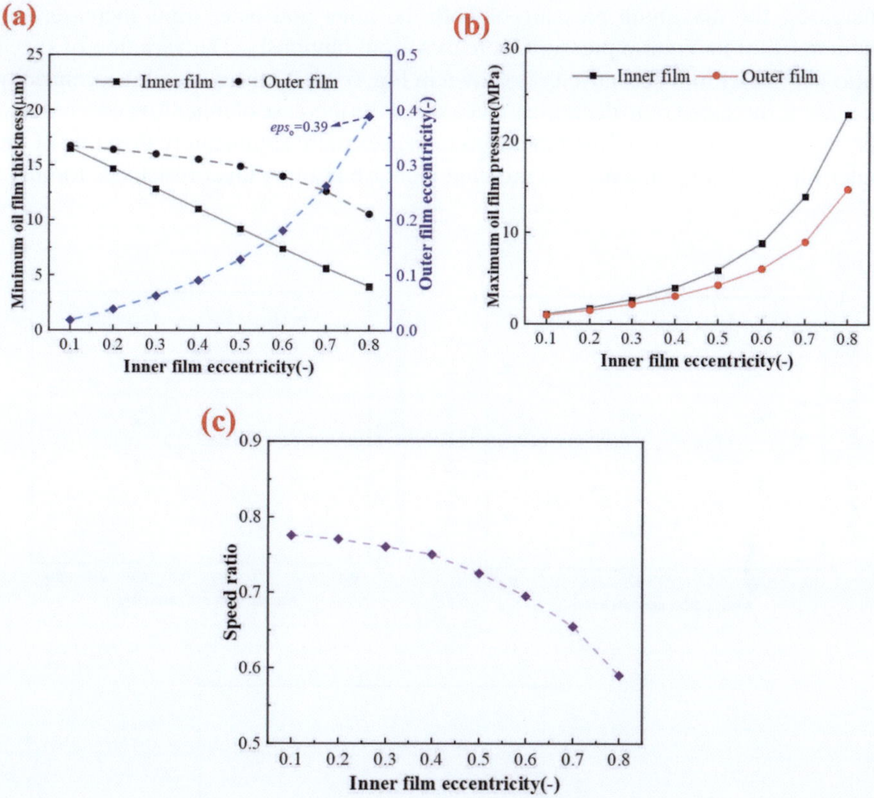

Fig. 6.8 The influence of inner eccentricity on the lubrication characteristics of the RFBP structure **a** Minimum oil film thickness and outer film eccentricity **b** Maximum oil film pressure **c** Speed ratio

6.4.1.4 The Influence of Outer Eccentricity on Lubrication Characteristics

The variation curve of the lubrication characteristics of the RFBP structure which change with the eccentricity of the outer oil film, is shown in Fig. 6.9. The variation curve of the minimum thickness of the inner/outer film with the outer film eccentricity is shown in Fig. 6.9a. And the variation curve of the inner film eccentricity with the outer film eccentricity is also shown in Fig. 6.9a. It can be seen that as the outer film eccentricity increases, the minimum thickness of both the inner and outer films decrease, and the decrease in the inner film is greater. In addition, it can be seen that when stable state is reached, the outer eccentricity is greater than the inner eccentricity, and the inner eccentricity increases with the outer eccentricity. When the outer eccentricity reaches 0.8, the inner eccentricity has reached 1.09. The variation curve of the maximum pressure of the inner/outer film with the inner film eccentricity is shown in Fig. 6.9b. It can be seen that as the outer film eccentricity increases, the maximum pressure of both the inner and outer films increase, and the inner film maximum pressure increases more obviously. The variation of speed ratio with outer film eccentricity is shown in Fig. 6.9d. As the outer film eccentricity increases, the speed ratio decreases. Because of the increase of inner film eccentricity, the maximum pressure of the inner film increases more significantly than that of the outer film, resulting in a more significant increase in inner layer resistance torque.

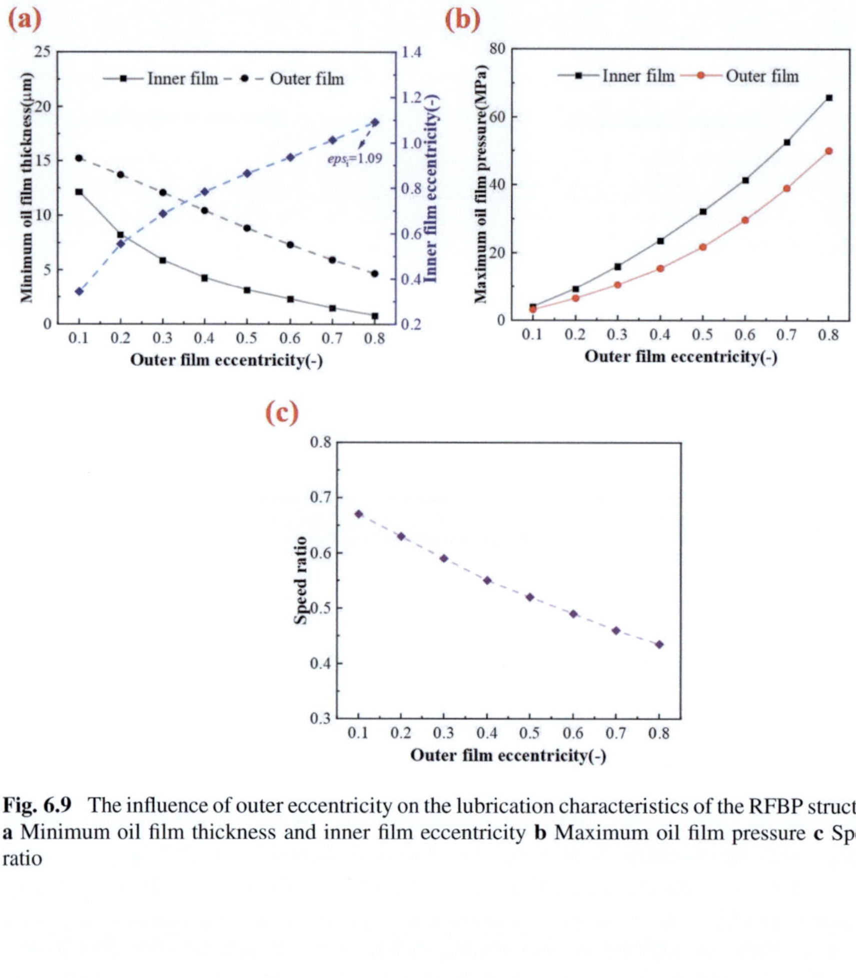

Fig. 6.9 The influence of outer eccentricity on the lubrication characteristics of the RFBP structure **a** Minimum oil film thickness and inner film eccentricity **b** Maximum oil film pressure **c** Speed ratio

6.4.2 Dynamic Characteristic Analysis of the RFBP Structure

6.4.2.1 The Influence of Roller Speed

The influence of roller speed on the dynamic parameters of the inner and outer films is shown in Fig. 6.10. The variation of stiffness coefficients of the inner and outer oil films with roller speed is shown in Fig. 6.10a, b. It can be seen that the main/cross stiffness coefficients of the inner and outer films decrease with the increase of roller speed. The main stiffness coefficients of the inner oil film are greater than that of the outer oil film, and the cross stiffness coefficients are smaller than that of the outer oil film. The variation of inner and outer oil film damping coefficients with roller speed is shown in Fig. 6.10c, d. It can be observed that as the roller speed increases, the main damping coefficients of the inner and outer films decrease, while the cross damping coefficients increase. Moreover, the main damping coefficients of the inner film is greater than that of the outer film, while the cross damping is smaller. In addition, with the increase of roller speed, the stiffness/damping coefficients of the inner and outer oil films decrease. The result is that when the speed increases, the bearing capacity of the inner and outer oil films increases, and the disturbance on the oil film pressure decreases.

6.4.2.2 The Influence of External Load

The influence of external load (F) on the dynamic characteristics of the inner and outer films is shown in Fig. 6.11. The variation of stiffness coefficients of the inner and outer oil films with the external load is shown in Fig. 6.11a, b. It can be seen that the main stiffness coefficients of the inner and outer oil films increase with the increase of load, while the cross stiffness decreases. In addition, when the external load increases, the main/cross stiffness coefficients of the inner film increase more obviously compared to the outer oil film, because of the inner layer structure is smaller compared to the outer layer, with weaker load-bearing capacity, and the disturbance has a more significant impact on the inner layer oil film pressure. The variation of the damping coefficients of the inner and outer oil films with the external load is shown in Fig. 6.11c, d. It can be seen that as the load increases, the main damping coefficients of both the inner and outer oil films increase, while the cross damping coefficients decrease. Moreover, the main/cross damping coefficients of the inner oil film increase more obviously compared to the outer oil film.

6.4.2.3 The Influence of Length-Diameter Ratio

The influence of length-diameter ratio (L/D) on the dynamic characteristics of the inner and outer film is shown in Fig. 6.12. The variation of stiffness coefficients of

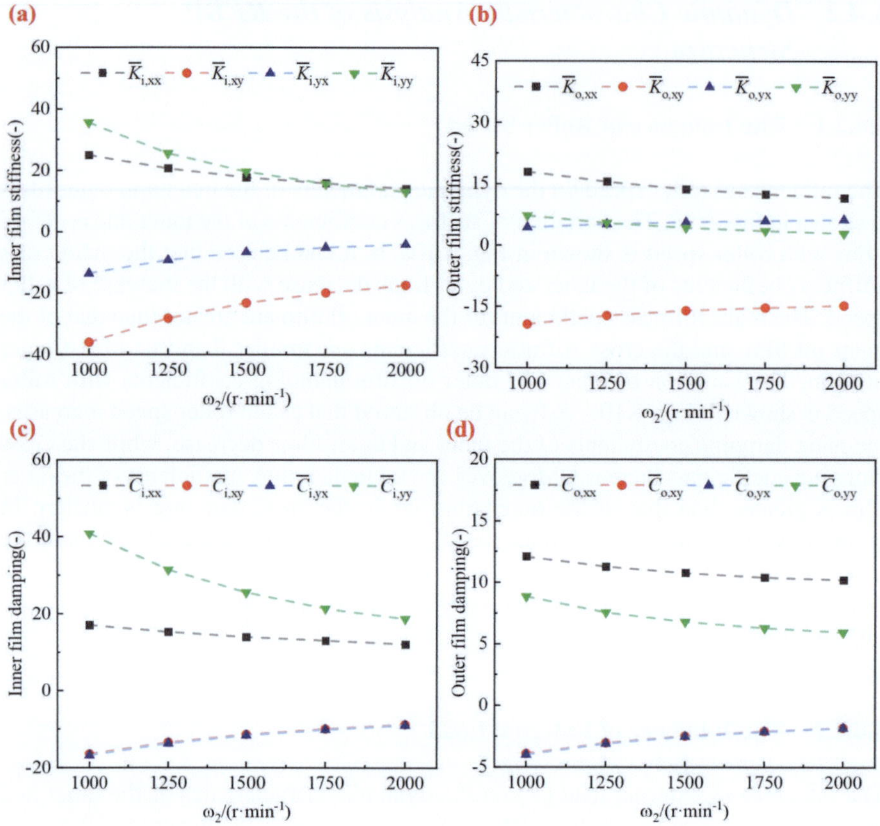

Fig. 6.10 The influence of roller speed on the dynamic characteristics of inner/outer oil film **a** Inner film stiffness **b** Outer film stiffness **c** Inner film damping **d** Outer film damping

the inner and outer oil films with length-diameter ratio is shown in Fig. 6.12a, b. It can be seen that as the length-diameter ratio increases, the main stiffness of the inner and outer oil films decreases, while the cross stiffness increases; Both the main/cross stiffness of the inner film decrease more significantly compared to the outer film. The variation of damping coefficients of the inner and outer films with length-diameter ratio is shown in Fig. 6.12c, d. It can be seen that as the length-diameter ratio increases, the main damping coefficients of the inner and outer films decrease, while the cross damping increases; The main damping coefficients of the inner film decrease more significantly compared to the outer film, while the cross damping coefficients of the inner oil film increase more significantly. In addition, as the aspect ratio increases, the stiffness and damping of the inner and outer oil films decrease.

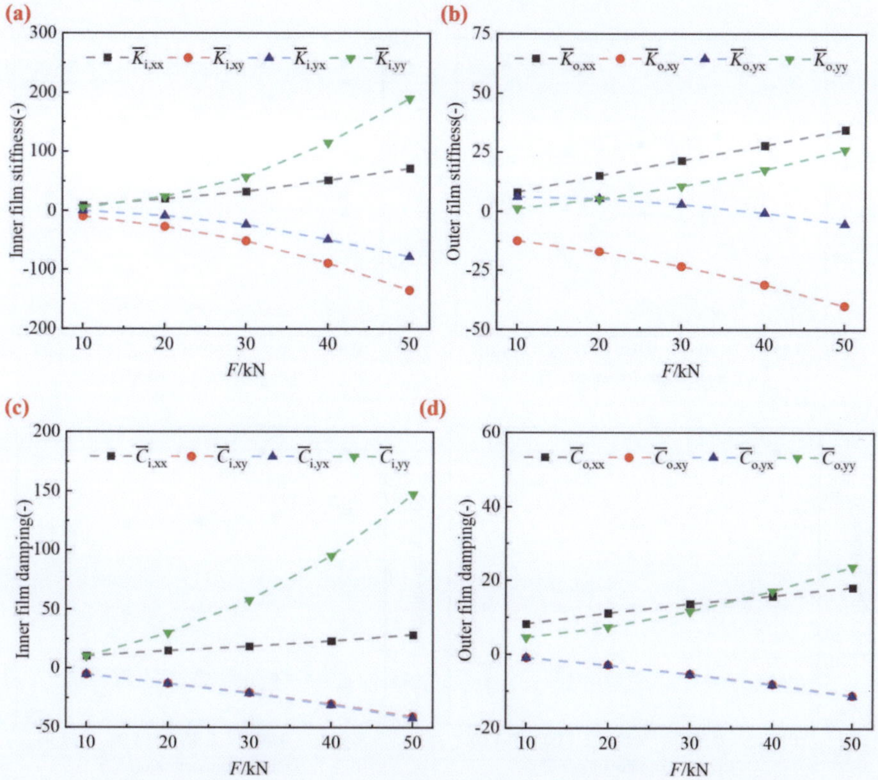

Fig. 6.11 The influence of external load on the dynamic characteristics of inner/outer oil film **a** Inner film stiffness **b** Outer film stiffness **c** Inner film damping **d** Outer film damping

6.4.2.4 The Influence of Clearance Ratio

The influence of the clearance ratio (c_i/c_o) on the dynamic characteristics of the inner and outer film is shown in Fig. 6.13. The clearance ratio is changed by fixing the outer clearance and changing the inner clearance. The variation of stiffness coefficients of the inner and outer oil films with the clearance ratio are shown in Fig. 6.13a, b. It can be seen that as the clearance ratio increases, the main stiffness coefficients of the inner film decrease, and the cross stiffness coefficients increase, while the stiffness coefficients of the outer film remain almost unchanged. In addition, when the clearance ratio increases, the stiffness coefficients of the inner oil film change more obviously, because of the decrease of load-bearing capacity of the inner oil film. The variation of damping coefficients of inner and outer films with clearance ratio is shown in Fig. 6.13c, d. It can be observed that as the clearance ratio increases, the main damping coefficients of the inner film increase, the cross damping coefficients decrease, and the damping of the outer film remains almost unchanged. In addition,

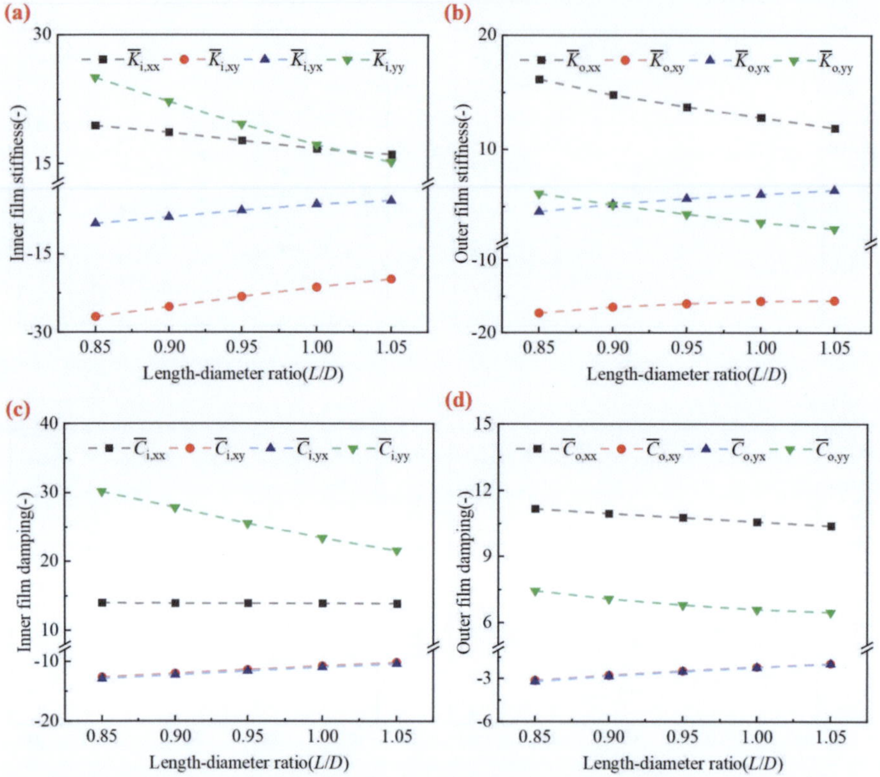

Fig. 6.12 The influence of length-diameter ratio on the dynamic characteristics of inner/outer oil film **a** Inner film stiffness **b** Outer film stiffness **c** Inner film damping **d** Outer film damping

the larger the coefficients ratio, the damping coefficients of the inner film change more obviously.

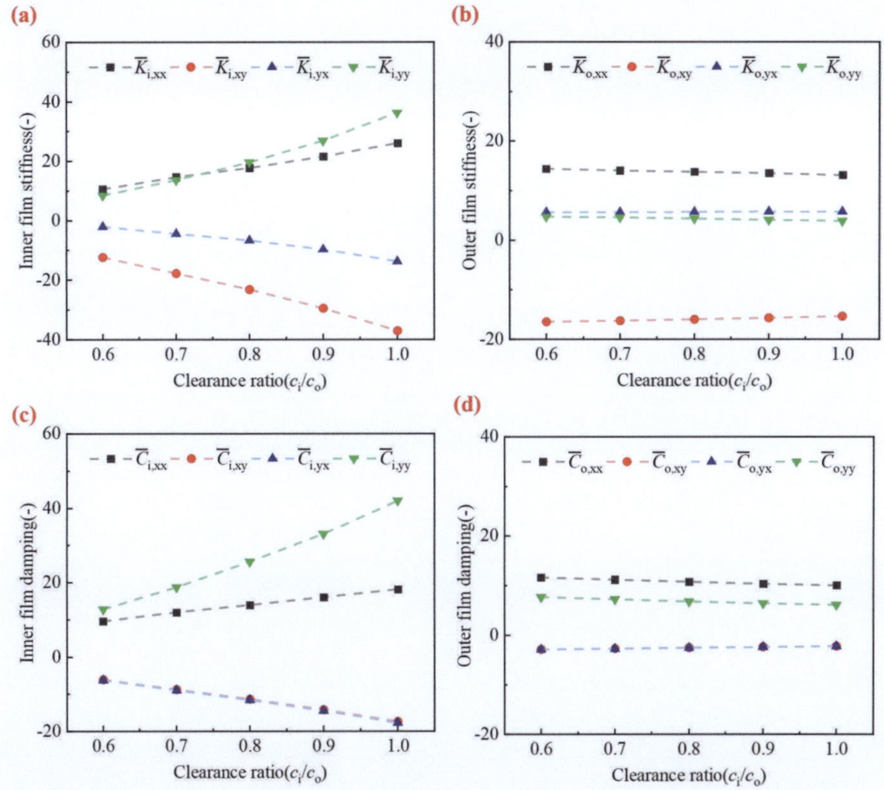

Fig. 6.13 The influence of clearance ratio on the dynamic characteristics of inner/outer oil film **a** Inner film stiffness **b** Outer film stiffness **c** Inner film damping **d** Outer film damping

References

Fang L (2016) Analysis of lubrication characteristics of floating ring bearings under poor oil conditions. Hebei University of Technology, Tianjin

Li YJ, Liang F, Zhou Y et al (2017) Numerical and experimental investigation on thermohydrodynamic performance of turbocharger rotor-bearing system. Appl Therm Eng 121:27–38

Patir N, Cheng HS (1979) Application of average flow model to lubrication between rough sliding surfaces. J Lubric Technol 101(2):220–229

Pei SY, Xu H, Yun M et al (2016) Effects of surface texture on the lubrication performance of the floating ring bearing. Tribol Int 102:143–153

Wang LK, Bin GF, Li XJ et al (2015) Effects of floating ring bearing manufacturing tolerance clearances on the dynamic characteristics for turbocharger. Chinese J Mech Eng 28(3):530–540

Dequan Y, Yujie D (2006) Boundary element analysis of the stability of misaligned floating ring bearings. Lubrication Sealing 1:38–40

Chapter 7
Transient Thermal Elastic Hydrodynamic Lubrication and Time-Varying Dynamic Characteristics of Double-Layer Oil Film Floating Bush

Abstract The roller-floating bush-pin (RFBP) structure is an important component in high-pressure fuel pumps of marine diesel engines. Considering the thermoelastic effect of the double-layer oil film structure and the time-varying working conditions of the fuel pump, a simulation model for the transient dynamic characteristics of the double-layer oil film structure and a test platform for the double-layer oil film lubrication system were established, verifying the accuracy of the simulation model. Through the comparative analysis of the dynamic parameters of the inner and outer oil films, the influences of working parameters and structural parameters on the dynamic characteristics of the structure were studied. The results show that, compared with the single-layer oil film structure, the double-layer oil film structure can significantly reduce the stiffness and damping of the structure. The dynamic characteristics of the internal oil film show greater sensitivity to changes in operating parameters and structural parameters. The influence of the outer oil film on the equivalent dynamic characteristics of the RFBP structure is more significant than that of the inner oil film. This work can provide useful guidance for the design of roller floating bushing pins in the oil supply mechanism.

Keywords Double-layer oil film structure · Stiffness · Damping · RFBP structure

7.1 Introduction

At present, scholars have conducted some research on bearings with double-layer oil film structures. The floating ring bearing in turbochargers is the most typical double-layer oil film structure. Many scholars have conducted research on the lubrication characteristics of the floating ring bearing structure. Some scholars have analyzed the lubrication performance of floating ring bearings with different structures (Soni et al. 2017a, 2017b; Pei et al.2016; Yang et al. 2023). Soni et al. (2017a, 2017b) studied the lubrication characteristics of non-circular floating ring bearings and found that compared with standard floating ring bearings, the non-circular structure can enhance the load-bearing capacity. Pei et al. (2016) found that surface texture can reduce the

B. Zhao et al., *Lubrication and Dynamic Characteristics Analysis Technology*,
https://doi.org/10.1007/978-981-95-2270-5_7

temperature rise of floating ring bearings; Yang et al. 2023 found that floating ring bearings with structural misalignment can improve operational stability. Wang et al. found that the vibration of turbochargers can be reduced by changing the combination of manufacturing clearances. Yang et al. (2023) found that the floating ring oil hole could reduce the temperature rise of end leakage. The relevant research mainly focuses on the influence of different structures on the lubrication characteristics. The lubrication model is slightly simplified, and the hydrodynamic lubrication model is mostly adopted. Some scholars have improved and revised the lubrication simulation model of floating ring bearings, taking into account more influencing factors in actual operation (Andrés et al. 2004, 2007; Liqiang et al. 2023; Chao et al. 2017; Novotny et al. 2018; Yang et al. 2023). Andrés et al. (2004, 2007) considered the influence of the oil-thermal effect and the heat transfer of the floating ring in the lubrication model of the floating ring bearing. Liqiang et al. (2023) studied the lubrication state of floating ring bearings under poor oil conditions and considered the cavitation effect when the oil lubrication was poor. Chao et al. (2017) established a steady-state lubrication model for floating ring bearings based on the Reynolds equation and the random rough model. Novotny et al. (2018) established an effective computational model for the rotor-floating ring bearing system, thereby solving the transient problem of turbocharger rotor dynamics. Yang et al. (2023) calculated the elastic deformation of the floating ring by analytical method and included it in the lubrication model. They found that the deformation of the inner surface of the floating ring was greater than that of the outer surface.

Furthermore, since the dynamic characteristics of the floating ring bearing play an important role in the stability of the turbocharger rotor-floating ring bearing system, researchers have also conducted a series of studies on the dynamic characteristics of the floating ring bearing (Zhao et al. 2017; Bin et al. 2018; Hong et al. 2009; Hong et al. 2011; Wang et al. 2017; Zhongliang et al. 2022). Zhao et al. (2017) established the hydrodynamic lubrication model of the floating ring bearing and conducted a preliminary analysis of the stiffness/damping coefficient of the inner and outer oil films by using the micro-disturbance method. Bin et al. (2018) established a finite element model of the turbocharger rotor system based on the stiffness/damping coefficient of the floating ring bearing. The research finds that the excessive cross stiffness of the bearing is the main cause of self-excited vibration. Hong et al. (2009, 2011) established the lubrication models of floating ring bearings with deep and shallow cavities and calculated the stiffness/damping coefficient of the floating ring bearing structure. On this basis, the influence of rotor stiffness and float ring mass on the stability of the float ring bearing rotor system was also discussed. Wang et al. (2017) analyzed the influence of current intensity on the stiffness coefficient of floating ring bearings with magnetic pole structure. Zhongliang et al. (2022) established a lubrication calculation model for floating ring bearings considering rough surfaces. On this basis, the influence of working condition parameters and structural parameters on the stiffness/damping coefficient of the bearing was systematically studied.

The above work mainly studies the floating ring bearing structure in the turbocharger, while this paper focuses on the roller-floating bushing-pin structure in the diesel engine fuel supply mechanism. There are significant differences between the two. Firstly, in terms of structure and working mode, when the roller-floating bushing-pin structure is in operation, the CAM drives the outer roller, and the outer roller drives the floating bushing to rotate. The roller-pin is a fixed component. In contrast, when the floating ring bearing structure is in operation, the rotation of the internal rotor drives the floating ring, while the external bearing is fixed. Furthermore, in terms of working conditions, the floating ring bearing structure usually operates under stable high-speed and light-load conditions, while the roller-floating bushing-pin structure typically operates under low-speed and heavy-load conditions. Therefore, the elastic deformation and mixed lubrication of the roller-floating bushing-pin structure are more obvious than those of the floating ring bearing structure. Furthermore, since the working state of the roller-floating bushing-pin structure is stable, most scholars adopt the steady-state model to study its lubrication state and dynamic characteristics. However, the roller-floating bushing-pin structure usually operates in a transient state, and its operating state is related to the dynamics and kinematics of the CAM mechanism of the high-pressure oil pump. At present, scholars mainly study the lubrication and dynamic characteristics of the roller-floating bushing-pin structure under steady-state conditions (El-Ghoul et al. 2019; Alakhramsing et al. 2018), ignoring the actual instantaneous conditions and simplifying the elastic deformation and roughness characteristics. Therefore, it is necessary to establish the transient lubrication model and dynamic characteristic model of the roller-floating bushing-pin structure, and comprehensively consider the dynamics and kinematics of the CAM mechanism.

To sum up, at present, the research on the lubrication characteristics of the roller-floating bushing-pin structure of Marine diesel engine fuel pumps is limited, and the research on the transient dynamic characteristics based on the lubrication characteristics is even rarer. Furthermore, the research on the double-layer oil film lubrication structure mainly focuses on the floating ring bearing structure in turbochargers. However, the roller-floating bushing-pin structure in this study is quite different from the floating ring bearing structure. Since the floating ring bearing usually operates under stable conditions, while the roller-floating bushing-pin structure in this work operates under transient conditions, Its working conditions are also significantly different from those of floating ring bearings. Due to the crucial role of the dynamic characteristics of the roller-floating bushing-pin structure in the vibration analysis of fuel pumps, it is of vital importance to establish a dynamic characteristic model under transient operating conditions. Therefore, considering the thermoelastic effect of the roller-floating bushing-pin structure, the cam-roller dynamic and kinematic models are coupled with the transient lubrication model of the roller-floating bushing-pin structure, and the dynamic characteristics of this structure are calculated by using the small disturbance method. Based on the above model, the lubrication characteristics and dynamic characteristics of the inner and outer oil films were comparatively studied. Meanwhile, the influences of working conditions and structural parameters on the lubrication characteristics and dynamic characteristics were also investigated. The research results can provide valuable guidance for the design optimization of rollers—floating bushings—pins.

7.2 Transient State Lubrication and Dynamic Characteristic Analysis Model Under the Coupling Effect of Double-Layer Oil Film

7.2.1 Cam-Roller Model

Figure 7.1 shows the model of the cam-roller utilized in the fuel supply mechanism of the diesel engine. The cam drives the roller to rotate, and the roller drives the floating bush to rotate while the roller pin is a fixed component. It can be seen that the force F on the roller mainly consists of three parts, including the fuel force of the plunger F_P, the inertia force of the roller follower F_N, and the plunger spring force F_C. M_p is the mass of plunger and other related component, and M_r is the mass of roller and followers. P is the contact point of the cam-roller. φ is the pressure angle of the cam mechanism. O_1 is the center of rotation of the cam, and O_c is the center of base circle. R_0 is the base circle radius, and R_1 is the cam curvature radius. R_2 is the outer radius of the roller. ω_1 and ω_2 are the speed of cam and roller, respectively. u_1 and u_2 are the surface velocity of cam and roller, respectively. Z and y are the vertical heights of the roller center and contact point P, respectively.

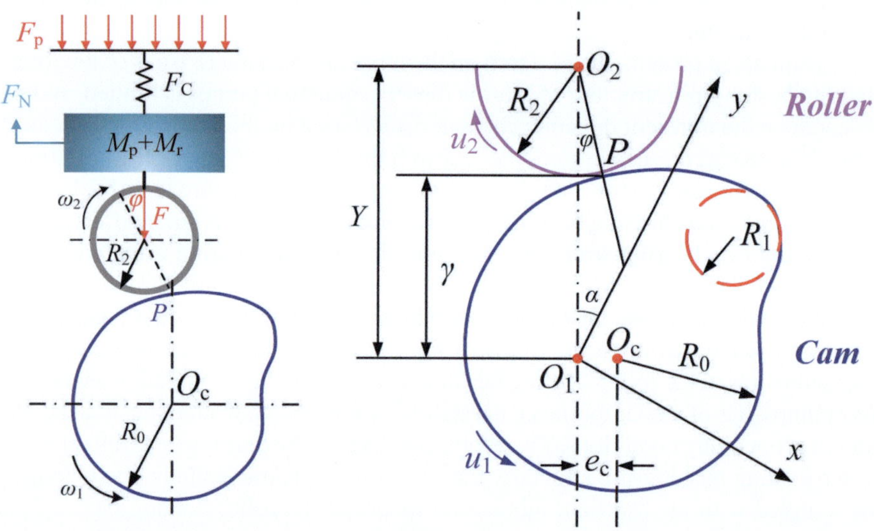

Fig. 7.1 Schematic diagram of cam-roller model in fuel pump

7.2.1.1 Cam-Roller Dynamic Model

The calculation formula for the time-varying load on the roller is as follows:

$$\begin{aligned} F \cdot \cos\varphi = & \underbrace{F_0 + c \cdot s_\alpha}_{F_C} + \underbrace{M\omega_1^2 \frac{\mathrm{d}^2 s_\alpha}{\mathrm{d}\alpha^2}}_{F_N} \\ & \underbrace{+ 0.25\pi \cdot d_z^2 \cdot \left[p_1 + (p_1 - p_2) \cdot s_\alpha / s_{\max}\right]}_{F_P} \end{aligned} \tag{7.1}$$

By giving the extreme value of the fuel pressure, F_P can be obtained by using Newton interpolation. The maximum and minimum extreme values of the fuel pressure are p_1 and p_2, respectively. M is the mass driven by cam, which is the sum of M_p and M_r. The pressure angle φ can be expressed as follow:

$$\varphi = \arctan \frac{|s_\alpha - e_c|}{s_\alpha + \sqrt{R_0^2 - e_c^2}} \tag{7.2}$$

7.2.1.2 Cam-Roller Kinematic Model

It can be seen from Fig. 7.1, there are geometric relationships in the cam mechanisms, as follow:

$$\begin{cases} Z = \gamma + R_2 \\ \gamma = s_\alpha + R_0 \end{cases} \tag{7.3}$$

For the curvature radius of the fuel pump cam R_1, the calculation formula is:

$$R_1 = \frac{\left[Y^2 + \left(\frac{\mathrm{d}Y}{\mathrm{d}\alpha}\right)^2\right]^{\frac{3}{2}}}{Y^2 + 2\left(\frac{\mathrm{d}Y}{\mathrm{d}\alpha}\right)^2 - Z\frac{\mathrm{d}^2 Y}{\mathrm{d}\alpha^2}} - R_2 \tag{7.4}$$

The derivative of the roller height (Y) in the y-direction with the cam angle is the roller velocity in the y-direction, while the second derivative of the roller height (Y) in the y-direction with the cam angle is the roller acceleration in y-direction. The velocity of cam (u_1) and roller (u_2) can be calculated as follow:

$$u_1 = \omega_1 \cdot R_{1-2},\ u_2 = \omega_2 \cdot R_2 \tag{7.5}$$

where $R_{1\text{-}2}$ is the comprehensive curvature radius of the cam-roller, and the calculation formula of $R_{1\text{-}2}$ is:

$$\frac{1}{R_{1-2}} = \frac{1}{R_1} + \frac{1}{R_2} \tag{7.6}$$

The roller velocity u_2 can be solved by the following Eq. (7.7):

$$u_2 = u_1 + \omega_1 R_2 \frac{\left(\frac{d\gamma}{d\alpha}\right)^2 - \gamma \frac{d^2\gamma}{d\alpha^2}}{\gamma^2 + \left(\frac{d\gamma}{d\alpha}\right)^2} \tag{7.7}$$

The roller speed (ω_2) can be obtained by combining Eqs. (7.6) and (7.7).

7.2.2 The Motion Trajectory of Roller and Floating Bush Calculation Model

Figure 7.2 presents the transient model of the RFBP structure. Figure 7.2a presents the middle plane of the RFBP structure. At first, the center of the floating bush and the roller were both located at point O_p, which is the center of the roller pin. Under the action of external loads, the floating bush center moves to Point O_f, while the roller center moves to Point O_2. Figure 7.2b presents the coordinate system of dynamic model. e_1 and e_3 represent the floating bush center eccentricity and the roller center eccentricity which is relative to the pin center. e_2 represents the eccentricity of the roller center relative to the floating bush center.

Equation (7.8) provides the equations for determining the radial and axial displacement trajectories of the floating bush.

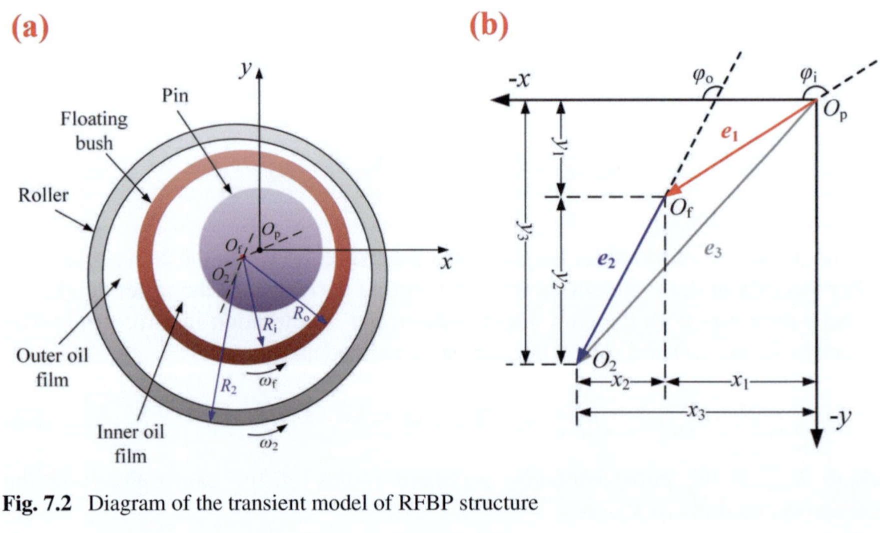

Fig. 7.2 Diagram of the transient model of RFBP structure

$$\begin{cases} m_1\ddot{x}_1 = P_{xi} - P_{xo} \\ m_1\ddot{y}_1 = P_{yi} - P_{yo} + m_1 g \end{cases}, \tag{7.8}$$

Equation (7.9) provides the equations for determining the radial and axial displacement trajectories of the roller.

$$\begin{cases} m_2\ddot{x}_2 = P_{xo} \\ m_2\ddot{y}_2 = F - P_{yo} \end{cases}, \tag{7.9}$$

Equation (7.10) provides the equations for determining the transient velocity of the floating bush.

$$I_1\ddot{\omega}_{\mathrm{f}} = \Gamma_{\mathrm{i}} - \Gamma_{\mathrm{o}}, \tag{7.10}$$

where Γ_{i} and Γ_{o} represent to the friction torque of the inner and outer layers of the floating bush respectively.

7.3 The Solution Procedure

Figure 7.3 shows the numerical analysis flow chart of transient dynamic characteristics calculation of the RFBP structure. The cam-roller model is calculated to obtain the transient operating conditions of the RFBP structure, and the results are taken as the input for the RFBP structure dynamic characteristics calculation model. The elasticity of the structure and the thermal effect of the lubrication oil were considered in the model. The calculation was completed when the motion trajectory of the roller and floating bush closed, respectively, within the cycle. The super relaxation iteration is used to accelerate the calculation speed, and the convergence accuracy of pressure calculation is 10–6. The displacement disturbances of the inner/outer oil films are both 1.0 μm. The velocity disturbances of the inner/outer oil films are set to 0.01 times the relative rotational speed. In addition, during the calculation process, the grids of inner/outer oil films are divided into 120 × 16 (circumference × axial).

7.4 Model Validation

In order to verify the reliability of the double-layer oil film transient lubrication model, an experimental platform was built to test the speed of the floating bush. The structural diagram of the experimental platform is shown in Fig. 7.4. The camshaft is driven to rotate by a servo motor (4.5 kW power), and the cam drives the roller to rotate. The RFBP structure is loaded with springs, and the diameter of the springs is 6.0 mm. The torque sensor (DYN-200) was used to measure the speed of the

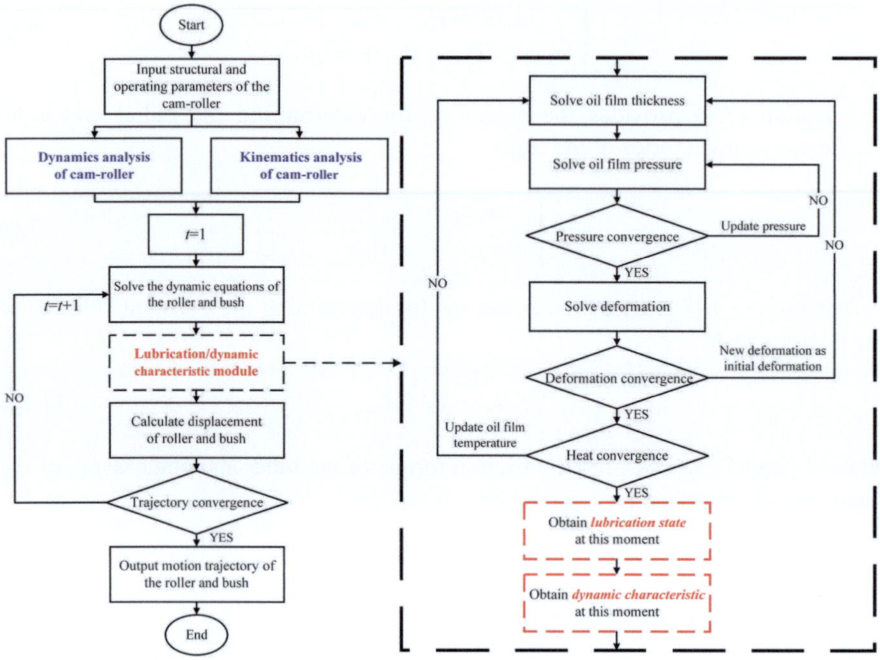

Fig. 7.3 Flow chart of numerical analysis for thermoelastic hydrodynamic lubrication of the RFBP structure

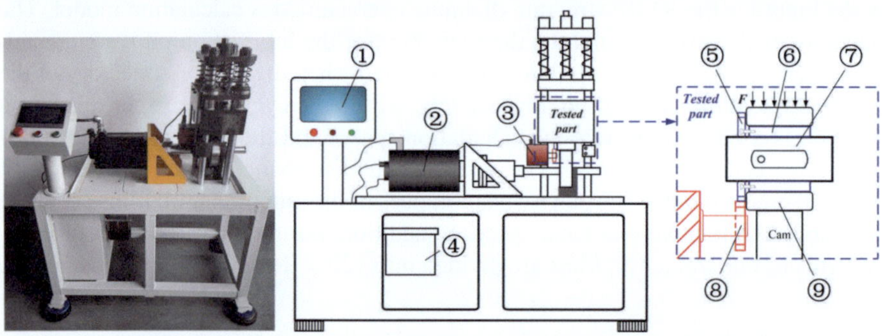

Fig. 7.4 Test bench structure diagram. (1) Display screen (2) Electric motor (3) Speed sensor (4) Oil tank (5) Floating bush gear (6) Floating bush (7) Roller pin (8) Sensor gear (9) Roller

floating bush. The testing method involved fixing a gear to the side of the floating bush and meshing it with the sensor gear. The gear transmission is used to output the time-varying speed of the floating bush. The cam profile and lift curve used in the experiment are shown in Fig. 7.5.

The comparison results are shown in Fig. 7.6. The comparison results of the average speed of the floating bush at different cam speeds are shown Fig. 7.6a. The

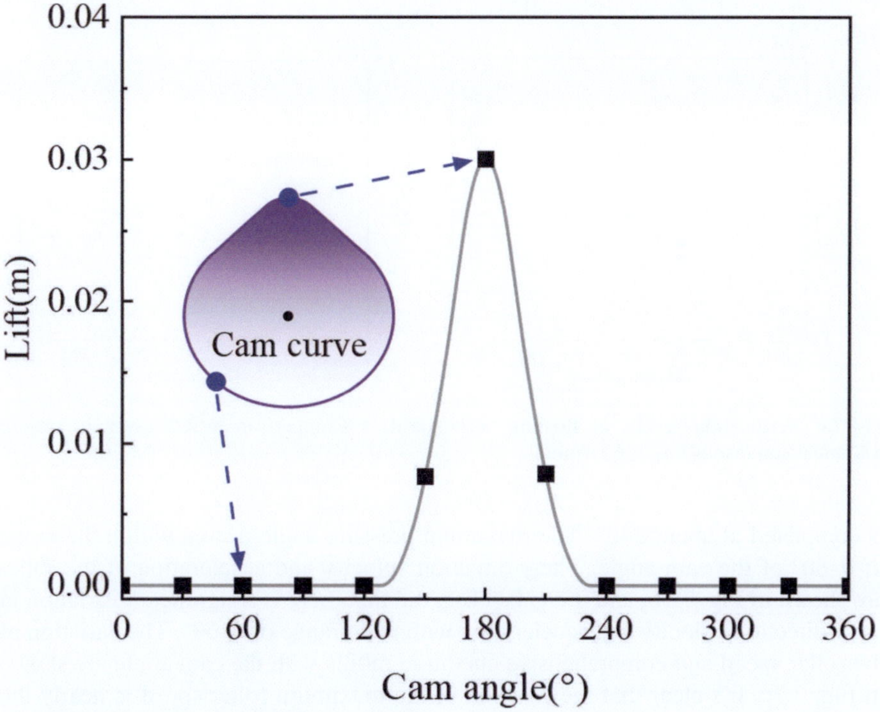

Fig. 7.5 Cam profile and lift curve in the test

average speed of the floating bush within the cycle is expressed by $\overline{\omega}_f$. The calculated results and tested results both increase with the cam speed, and the deviation is within 6%. Figure 7.6b shows that the instant speed comparison results of the floating bush. The calculated results of the floating bush speed variation trend with the cam angle are closely matches the tested results, and the numerical deviation of no more than 10%. In addition, the tested results are larger than the calculated results due to the oil holes around the floating bush in the test, a factor which is not considered in the simulation model. The accuracy of the simulation model's calculation results is proved by the verification results.

7.5 Result Analysis

7.5.1 Analysis of Cam-Roller Model

The result analysis of cam roller is shown in Fig. 7.7. The variation curves of lift and pressure angle of the cam mechanism with the cam angle are shown in Fig. 7.7a. It can be seen that the highest lift occurs at about 60° of cam angle, and the return section

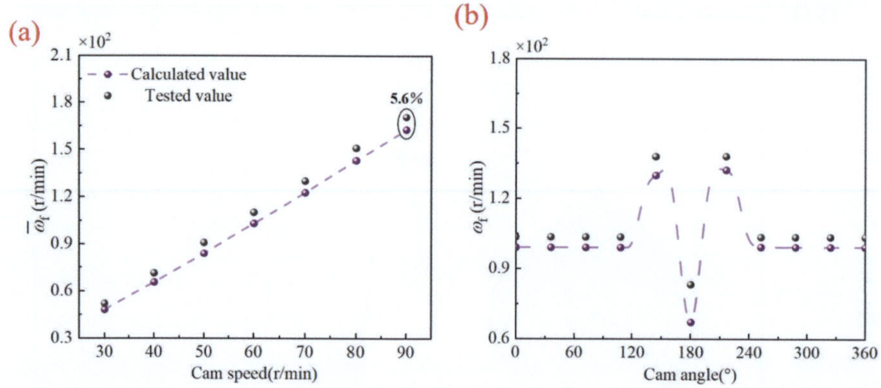

Fig. 7.6 Verification results of floating bush speed. **a** Comparison results (average speed); **b** Comparison results ($\omega_1 = 60$ r/min)

is completed at about 240°. The maximum pressure angle occurs within the range of 0–60° of the cam angle. The *y*-direction velocity and acceleration of the roller are shown in Fig. 7.7b, and it can be observed that there is a significant variation in the *y*-direction velocity and acceleration within the range of 0–60°. The variation of the roller speed and comprehensive curvature radius with the cam angle are shown in Fig. 7.7c. It's clear that the position of the maximum roller speed is nearly the same as the position of the maximum comprehensive curvature radius. The variation of fuel pressure and the sum of inertia and spring force with cam angle are shown in Fig. 7.7d. It is clear that the fuel pressure value is much higher than the sum of inertia and spring forces, indicating that the main load on the roller comes from the fuel pressure. When the cam angle is about 60°, the fuel pressure reaches the peak, about 39.36 MPa.

7.5.2 Analysis of the RFBP Structure Lubrication/Dynamic Characteristics

The change of the lubrication characteristics of the RFBP structure with cam angle is shown in Fig. 7.8. The motion trajectory of the roller and floating bush is shown in Fig. 7.8a. The centers of the roller and floating bush motion trajectories are both the roller pin center. The *x*-direction and *y*-direction components of the floating bush displacement correspond to x_1 and y_1, respectively, and *x*-direction and *y*-direction components of the roller displacement correspond to x_3 and y_3, respectively. It can be seen that due to the vertical downward direction of the external load of the roller, the y-direction motion component of the roller and the floating bush was negative at the beginning. When the external load is lower, both the roller and the floating bush gradually return to their initial position. Moreover, the range of motion of the roller

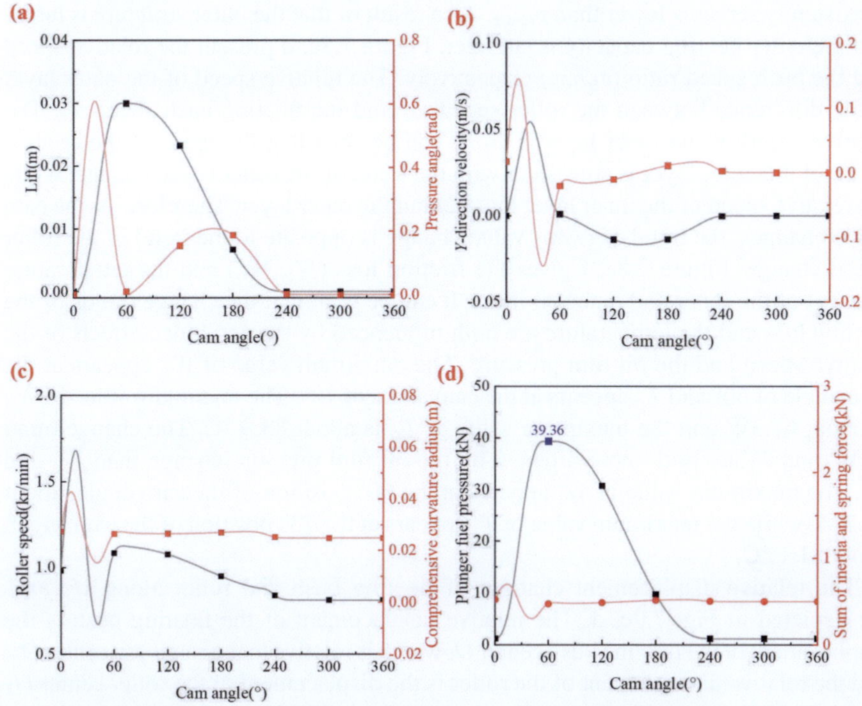

Fig. 7.7 Verification results of the dynamic characteristic parameters. **a** The variation of the lift and pressure angle with cam angle; **b** The variation of *y*-direction velocity and acceleration with cam angle; **c** The variation of comprehensive curvature radius and roller speed with cam angle; **d** The variation of fuel pressure and the sum of inertia and spring force with cam angle

is relatively large, while the range of motion of the floating bush is relatively small. This is because the outer oil film force acting on the floating bush is smaller than the external load on the roller due to the motion of the roller. Figure 7.8 presents the maximum oil film pressure of the inner and outer layers ($p_{i,max}$, $p_{o,max}$) of the RFBP structure with the cam angle. And the figure also shows the oil film pressure distribution at the position of the maximum value of the curve. During the cam angle's transition from 0° to 360°, both $p_{i,max}$ and $p_{o,max}$ experience an initial increase, followed by a decrease, and ultimately remains unchanged. The variation trend of $p_{i,max}$ and $p_{o,max}$ with cam angle is similar to that of the external load on the roller. The position where both the maximum value of $p_{i,max}$ and $p_{o,max}$ occur lags behind the position of the maximum load, and the position where the maximum value of $p_{i,max}$ occurs lags behind the outer layer, due to the effect of squeezing. The maximum value of $p_{i,max}$ occurs at the cam angle of about 91°, which is about 68.3 MPa, and the maximum value of $p_{o,max}$ occurs at the cam angle of about 81°, which is about 51.9 MPa. In addition, it can be seen that the pressure distribution of the inner oil film is more concentrated than that of the outer film, and for every cam angle, $p_{o,max}$

consistently remains lower than $p_{i,max}$. As a result of that the outer structure is larger and the outer bearing capacity is stronger. Figure 7.8c, d present the relative speed and the bush speed ratio (ω_r/ω_2) respectively. The relative speed of the outer layer is the difference between the roller speed ω_2 and the floating bush speed ω_r. The relative speed of the inner layer is ω_r. It can be seen that the trend of the relative speed of the outer layer is consistent with the trend of the roller speed change, while the relative speed of the inner layer lags behind the outer layer. Therefore, as the cam angle changes, the trend of ω_r/ω_2 value change is opposite to the trend of the roller speed change. Figure 7.8e, f gives the friction loss (W_i, W_o) and the temperature (T_i, T_o) of the inner and outer oil films. It can be seen that the change trends of the friction loss and the temperature are both influenced by the combined effects of the relative speed and the oil film pressure. The maximum value of W_o appears at the cam angle of 60° and T_o appears at the cam angle of 10°. The maximum value of W_o is about 67.3W and the maximum value of T_o is about 83.9 °C. The change trend of W_i and T_i are both more affected by the oil film pressure change than W_o and T_o. The maximum value of W_i appears at the 30° position of the cam angle, about 54.7 W, while the maximum value of T_i appears at the 73° position of the cam angle, about 81.1 °C.

The relative displacement changes of floating bush and roller along *x*/*y* axes are depicted in Fig. 7.9c, d. The relative displacement of the floating bush is the displacement of the floating bush center O_f which is relative to the roller pin center O_p, and the relative displacement of the roller is the displacement of the roller center O_2 which is relative to the floating bush center O_f. It can be observed that the maximum distance between O_f and O_p in the *x*-direction occurs at a cam angle of about 300°, which is about 12.2 μm. The maximum distance between O_f and O_p in the *y*-direction occurs at a cam angle of about 120°, which is about 12.5 μm. In addition, the peak distance in the *x*-direction between O_2 and O_f occurs at a cam angle of about 210°, which is about 11.2 μm. The peak distance in the *y*-direction between O_2 and O_f occurs at a cam angle of about 60°, which is about 12.3 μm.

Figure 7.10 presents the evolution of stiffness/damping coefficients as the cam angle changes. Since the main stiffness/damping coefficients have a greater impact on the dynamic performance of equipment structure compared to the cross stiffness/damping coefficients, and the variation law with eccentricity is similar, the work uses main stiffness/damping coefficients to perform a parametric analysis. The variation curves of the main stiffness with the cam angle are shown in Fig. 7.10a–c. The peak values of the primary stiffness along the *x*- and *y*-axes are about 43.3 and 86.8, attained at cam angles about 270° and 150°, respectively. The main stiffness variations along the *x*-axis during cam rotation correlate with the oil film's relative displacement changes in the corresponding direction, while an analogous relationship is observed for the *y*-axis components. The phenomenon conforms to the linear relationship between main stiffness and oil film eccentricity. The outer oil film exhibits lower main stiffness compared to its inner counterpart, while the equivalent main stiffness of the RFBP structure is smaller than the main stiffness of both the inner and outer films. The outer oil film's main stiffness has a more pronounced influence on the equivalent stiffness compared to inner counterpart. The variation curves

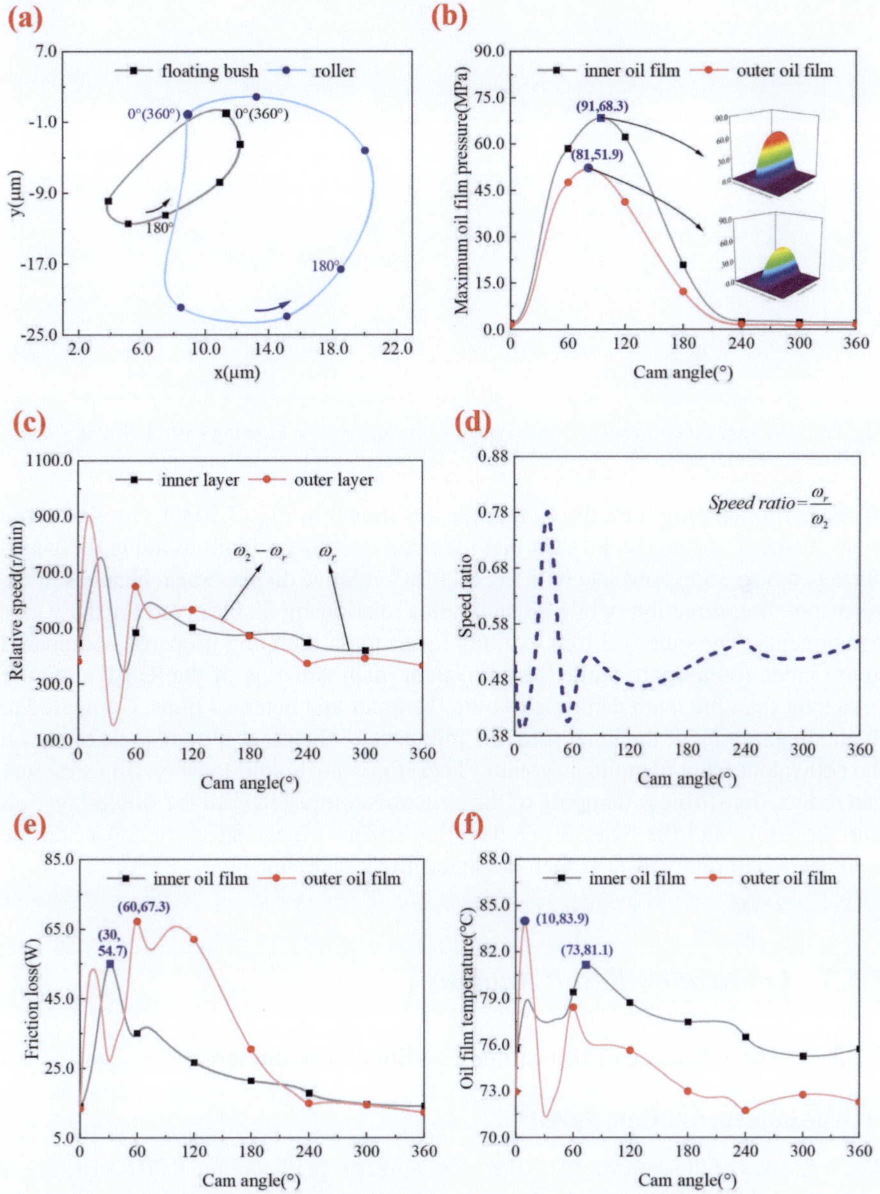

Fig. 7.8 The change of the lubrication properties with cam angle. **a** Movement trajectory of roller and floating bush; **b** The variation of $p_{i,max}$ and $p_{o,max}$ with cam angle and the oil film pressure distribution at the position of the maximum value the curve; **c** The variation of relative speed with cam angle; **d** The variation of speed ratio (ω_r/ω_2) with cam angle; **e** The variation of W_i and W_o with cam angle; **f** The variation of T_i and T_o with cam angle

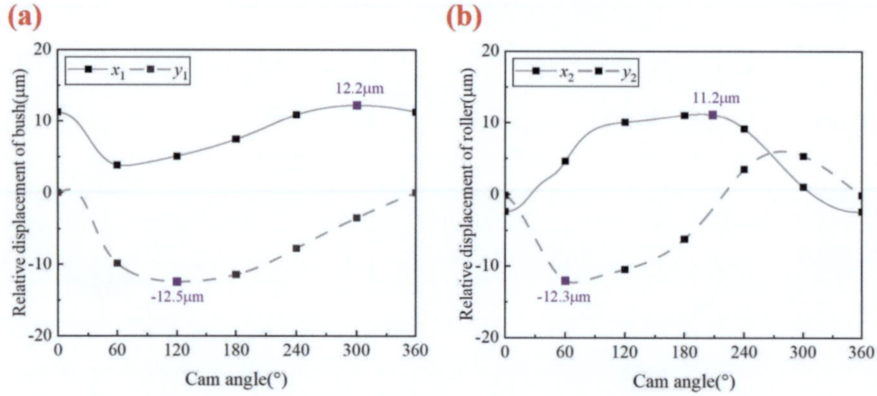

Fig. 7.9 The variation of relative displacement with cam angle **a** Floating bush; **b** Roller

of the main damping with the cam angle are shown in Fig. 7.10d–f. Similar to the main stiffness, it can also be seen that the main damping variations along the *x*-axis during cam rotation correlate with the oil film's relative displacement changes in the corresponding direction, while an analogous relationship is observed for the *y*-axis components. The outer oil film exhibits lower main damping properties compared to the inner counterpart, while the equivalent main damping of the RFBP structure is smaller than the main damping of both the inner and outer oil films. Compared to the main damping of the inner film, the influence of the outer film main damping on the equivalent main damping is greater. Therefore, the double-layer oil film structure can reduce the stiffness/damping of the structure contrasted with the single-layer oil film structure, and the outer film's damping/stiffness has a stronger impact on the equivalent damping/stiffness than the inner film's damping.

7.5.3 Lubrication Result Analysis

7.5.3.1 The Influence of Operating Conditions Parameters

(1) The influence of Cam Speed

The influence of the cam speed on the lubrication properties of the RFBP structure is given in Fig. 7.11. Figure 7.11a, b present the motion trajectory of the floating bush and roller expressed by eccentricity under different cam speed. The eccentricity of the floating bush is the ratio of e_1 to c_i, and the eccentricity of the roller is the ratio of e_3 to $(c_o + c_i)$. It can be seen that the motion trajectory of the floating bush and roller is mostly within the range of 210°–270° in the circumferential direction. As the cam speed increases, the eccentricity of the floating bush and roller both decreases, and the motion trajectory shifts towards the right. The reason is that as the cam speed

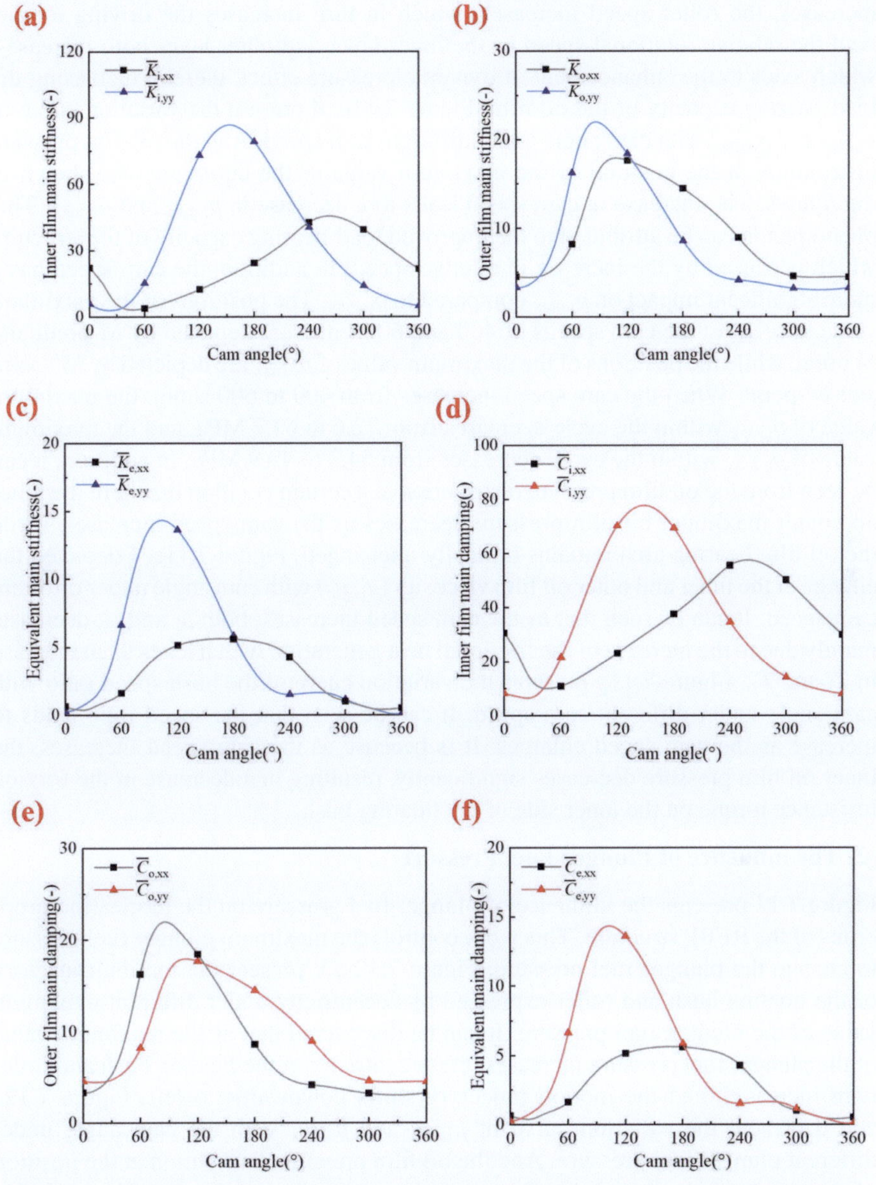

Fig. 7.10 The variation of stiffness/damping coefficients with cam angle. **a** The variation of inner film main stiffness with cam angle; **b** The variation of outer film main stiffness with cam angle; **c** The variation of equivalent main stiffness with cam angle; **d** The variation of inner film main damping with cam angle; **e** The variation of outer film main damping with cam angle; **f** The variation of equivalent main damping with cam angle

increases, the roller speed increases, which in turn increases the driving torque. And the relative rotational speed of the inner layer and outer layer both increases, which leads to the enhancement of dynamic pressure effect, thereby increasing the load-bearing capacity of the oil film. Figure 7.11c, d present the variation curve of $p_{i,max}$ and $p_{o,max}$ with cam angle under different cam speed. And the oil film pressure distribution at the position of the maximum value of the curve are also shown in the figures. The increase in cam speed leads to a decrease in $p_{i,max}$ and $p_{o,max}$. This phenomenon can be attributed to the improved load-bearing capacity of the structure which is caused by the increase in relative speed. In addition, the cam speed has a more significant impact on $p_{i,max}$ compared to $p_{o,max}$. The positions of the maximum value of $p_{i,max}$ at the cam speeds of 400 and 600 r/min are depicted by M-point and N-point, while the positions of the maximum value of $p_{o,max}$ are depicted by M′-point and N′-point. When the cam speed increases from 400 to 600 r/min, the maximum value of $p_{i,max}$ within the cycle decreases from 72.6 to 64.7 MPa, and the maximum value of $p_{o,max}$ within the cycle decreases from 54.3 to 49.9 MPa. In addition, it can be seen from the oil film pressure distribution at a certain position that both the inner and outer maximum oil film pressure decreases as the cam speed increases, while the oil film bearing area remains basically unchanged. Figure 7.11e, f depicted the change of the inner and outer oil film viscosity (μ_i, μ_o) with cam angle under different cam speed. It can be seen that as the cam speed increases, both μ_i and μ_o decrease, mainly due to the increase of the frictional heat generation which leads to an increase in T_i and T_o. Figure 7.11g presents the variation curve of the bush speed ratio with cam angle under different cam speed. It can be seen that the speed ratio tends to increase as the cam speed enlarges. It is because as the cam speed increases, the inner oil film pressure decreases significantly, resulting in a decrease in the friction resistance torque on the inner side of the floating bush.

(2) The influence of Plunger Fuel Pressure

Figure 7.12 presents the influence of plunger fuel pressure on the lubrication properties of the RFBP structure. This work controls the maximum plunger fuel pressure to change the plunger fuel pressure. Figure 7.12a, b present the motion trajectory of the floating bush and roller expressed by eccentricity under different maximum value of the plunger fuel pressure. It can be discovered that as the maximum value of the plunger fuel pressure increases, the eccentricity of the floating bush and roller both increases, and the motion trajectory shifts downwards to left. Figure 7.12c and d present the variation curve of $p_{i,max}$ and $p_{o,max}$ with the cam angle under different plunger fuel pressure. And the oil film pressure distribution at the position of the maximum value of the curve are also shown in the figures. The positions of the maximum value of $p_{i,max}$ at the maximum plunger fuel pressure of 50.0 and 60.0 MPa are depicted by M-point and N-point, while the positions of the maximum value of $p_{o,max}$ are depicted by M′-point and N′-point. It can be discovered that as the maximum value of the plunger fuel pressure increases, both $p_{i,max}$ and $p_{o,max}$ increase. When the maximum value of the plunger fuel pressure increases from 50.0 MPa to 60.0 MPa, the maximum value of $p_{i,max}$ within the cycle increases from 60.1 MPa to 77.0 MPa, and the maximum value of $p_{o,max}$ within the cycle increases from 46.1

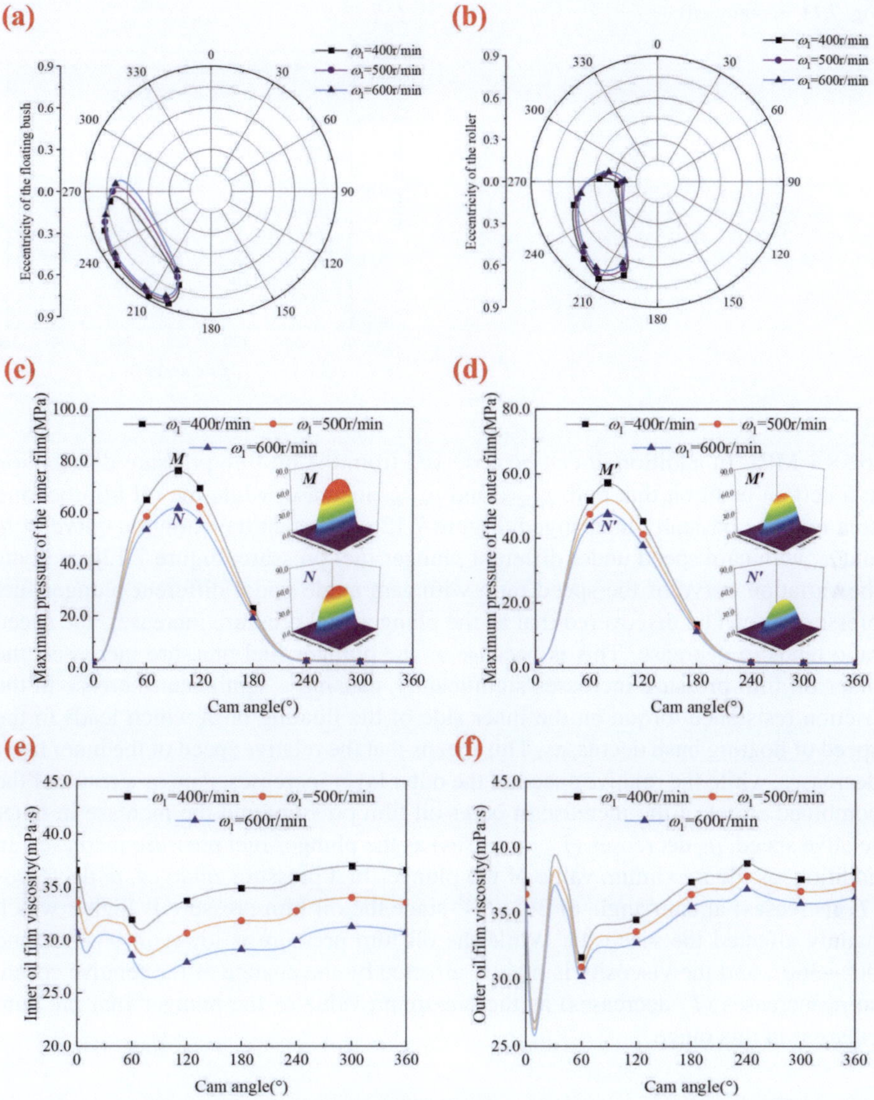

Fig. 7.11 The influence of cam speed on the lubrication properties of the RFBP structure. **a** The motion trajectory of the floating bush; **b** The motion trajectory of the roller; **c** The variation of $p_{i,max}$ with cam angle and the oil film pressure distribution at the position of the maximum value of $p_{i,max}$; **d** The variation of $p_{o,max}$ with cam angle and the oil film pressure distribution at the position of the maximum value of $p_{o,max}$; **e** The variation of μ_i with cam angle; **f** The variation of μ_o with cam angle under different cam speed; **g** The variation of speed ratio (ω_r/ω_2) with cam angle

Fig. 7.11 (continued)

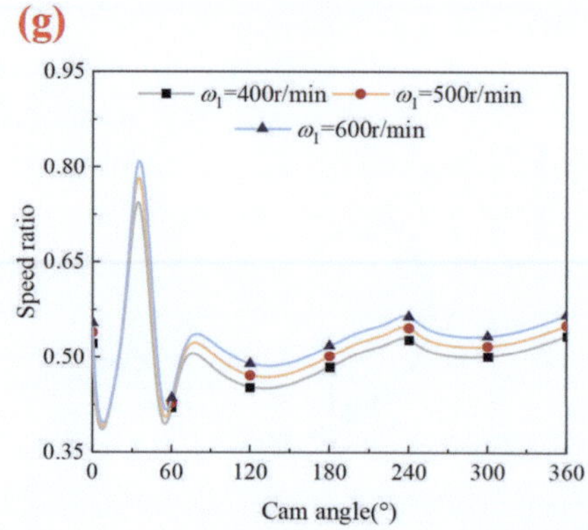

to 58.1 MPa. In addition, it can be observed from the oil film pressure distribution at a certain position that both $p_{i,max}$ and $p_{o,max}$ increase, while the oil film bearing area remains basically unchanged. Figure 7.12e, f present the variation curve of μ_i and μ_o with cam speed under different plunger fuel pressure. Figure 7.12g presents the variation curve of the speed ratio with cam angle under different plunger fuel pressure. It can be discovered that as the plunger fuel pressure increases, the speed ratio tends to decrease. This is because as the plunger fuel pressure increases, the inner oil film pressure increases significantly, causing a significant increase in the friction resistance torque on the inner side of the floating bush which leads to the speed of floating bush decreases. This means that the relative speed of the inner layer decreases, while the relative speed of the outer layer increases. And as a result of the combined effect of the increase in outer oil film pressure and the increase in outer relative speed, μ_o decreases (T_o increases) as the plunger fuel pressure increases. In addition, as the maximum value of the plunger fuel pressure enlarges, μ_i decreases (T_i increases) at cam angle of 30°–90° since the oil film pressure is higher which mainly affected the viscosity. While the oil film pressure is lower in 0°–30° and 90°–360°, and the viscosity is mainly affected by the change of the relative speed, so μ_i increases (T_i decreases) as the maximum value of the plunger fuel pressure enlarges in this range.

7.5.3.2 The Influence of Structural Parameters

(1) The Influence of Length-Diameter Ratios

Figure 7.13 presents the influence of length-diameter ratios (L/D) on the lubrication properties of the RFBP structure. Figure 7.13a, b present the variation curve of $p_{i,max}$ and $p_{o,max}$ with the cam angle under different L/D value. And the oil film pressure distribution at the position of the maximum value of the curve are also shown in the figures. The positions of the maximum value of $p_{i,max}$ at the L/D value of 0.85 and 1.05 are depicted by M-point and N-point, while the positions of the maximum

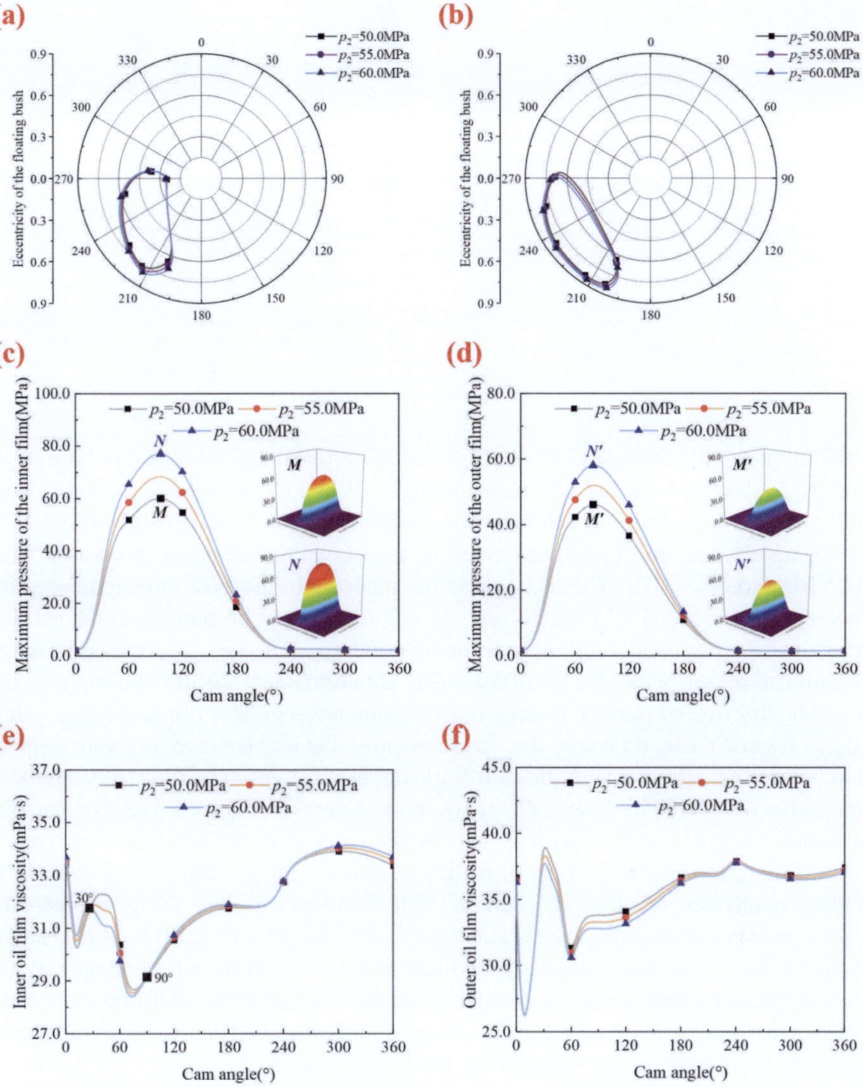

Fig. 7.12 The influence of plunger fuel pressure on the lubrication properties of the RFBP structure. **a** The motion trajectory of the floating bush; **b** The motion trajectory of the roller; **c** The variation of $p_{i,max}$ with cam angle and the oil film pressure distribution at the position of the maximum value of $p_{i,max}$; **d** The variation of $p_{o,max}$ with cam angle and the oil film pressure distribution at the position of the maximum value of $p_{o,max}$; **e** The variation of μ_i with cam angle; **f** The variation of μ_o with cam angle; **g** The variation of speed ratio (ω_r/ω_2) with cam angle

Fig. 7.12 (continued)

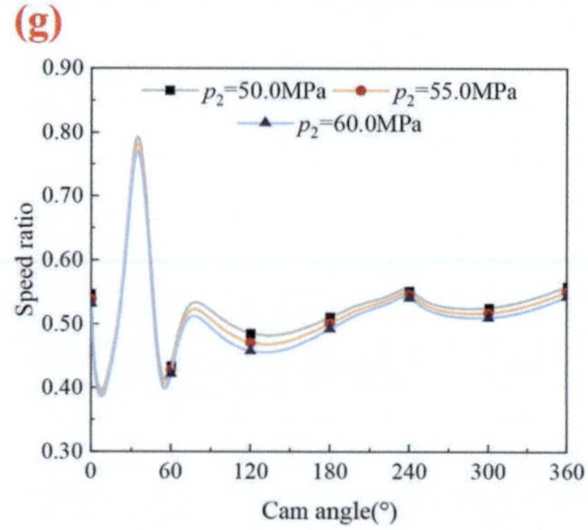

value of $p_{o,max}$ are depicted by M′-point and N′-point. It can be seen that both $p_{i,max}$ and $p_{o,max}$ decreases as the *L*/*D* value enlarges. When the *L*/*D* value increases from 0.85 to 1.05, the maximum value of $p_{i,max}$ within the cycle decreases from 83.5 MPa to 57.2 MPa, and the maximum value of $p_{o,max}$ within the cycle decreases from 62.5 MPa to 44.3 MPa. This is because the increase in the *L*/*D* value enhances the load-bearing capacity of the structure. In addition, it can be seen from the oil film pressure distribution at a certain position that both $p_{i,max}$ and $p_{o,max}$ decrease as the *L*/*D* value increases, while the oil film bearing area remains basically unchanged. And it can be discovered that the change in *L*/*D* value has a greater impact on $p_{i,max}$ than $p_{o,max}$. Figure 7.13c, d present the variation curve the end leakage flow rate of inner and outer films (Q_i, Q_o) with the cam angle under different *L*/*D* value. It can be seen that as the *L*/*D* value increases, Q_i and Q_o both decreases. Figure 7.13e, f present the variation curve of μ_i and μ_o with the cam angle under different *L*/*D* value. As a result of the increase of the *L*/*D* value, it is difficult for the lubrication oil end to leak, T_i and T_o both rises, resulting in μ_i and μ_o both decreases. Figure 7.13g indicates the variation curve of the bush speed ratio with the cam angle under different *L*/*D* value. It can be discovered that the speed ratio increases when the *L*/*D* value enlarges. When the *L*/*D* value increases, due to the significant decrease in inner oil film pressure, the friction resistance torque on the inner side of the floating bush decreases significantly, resulting in an increase in the speed of the floating bush.

(2) The influence of Clearance Ratios

The effect of clearance ratios (c_i/c_o) on the lubrication properties of the RFBP structure is shown in Fig. 7.14. This work controls the clearance ratio by fixing the outer clearance and changing the inner clearance. Figure 7.14a, b present the variation curve of the $p_{i,max}$ and $p_{o,max}$ with the cam angle under different c_i/c_o value. And the oil film pressure distribution at the position of the maximum value of the curve are also shown in the figures. The positions of the maximum value of $p_{i,max}$ at the c_i/c_o value of 0.7 and 0.9 are depicted by M-point and N-point, while the positions of the maximum value of $p_{o,max}$ are depicted by M′-point and N′-point. It can be seen that as the c_i/c_o value increases, $p_{i,max}$ increases, and with the growth of the clearance ratios

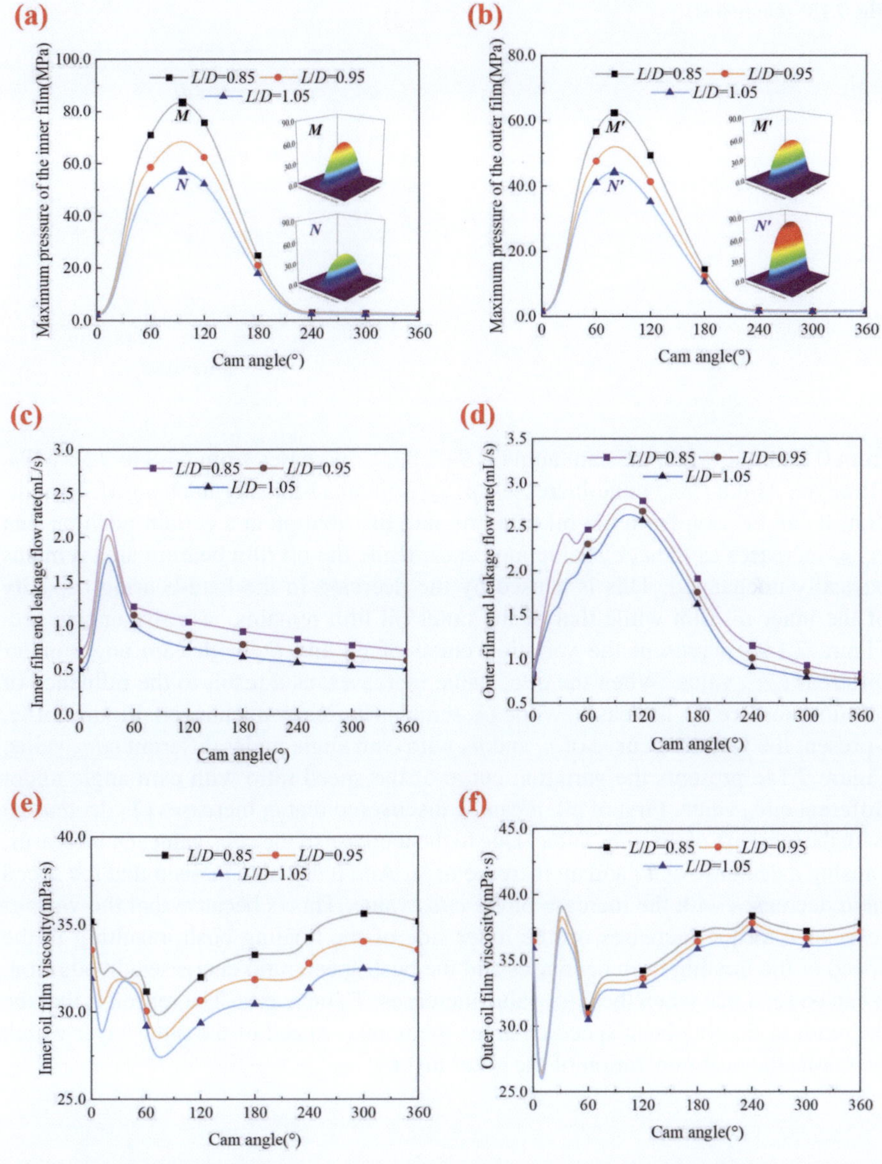

Fig. 7.13 The influence of *L*/*D* values on the lubrication properties of the RFBP structure. **a** The variation of $p_{i,max}$ with cam angle and the oil film pressure distribution at the position of the maximum value of $p_{i,max}$; **b** The variation of $p_{o,max}$ with cam angle and the oil film pressure distribution at the position of the maximum value of $p_{o,max}$; **c** The variation of Q_i with cam angle; **d** The variation of Q_o with cam angle; **e** The variation of μ_i with cam angle; **f** The variation of μ_o with cam angle; **g** The variation of speed ratio (ω_r/ω_2) with cam angle

Fig. 7.13 (continued)

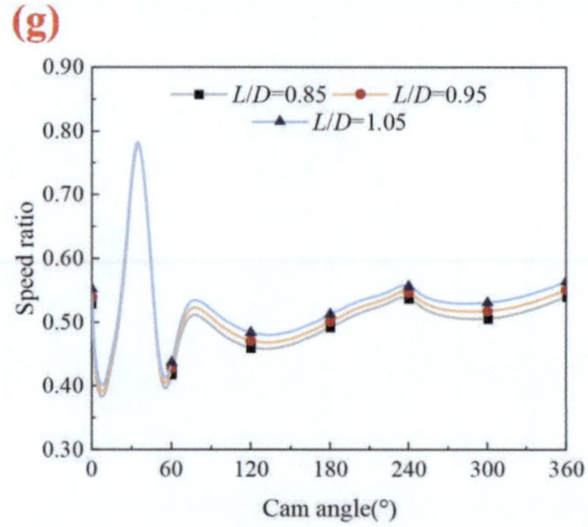

from 0.7 to 0.9, when the cam angle is 91°, $p_{i,max}$ increases from 63.8 to 72.9 MPa. However, as the c_i/c_o value increases, $p_{o,max}$ remains basically unchanged. In addition, it can be seen from the oil film pressure distribution at a certain position that $p_{i,max}$ increases as the c_i/c_o value increases, while the oil film bearing area remains basically unchanged. This is caused by the decrease in the load-bearing capacity of the inner oil film while that of the outer oil film remains basically unchanged. Figure 7.14c, d present the variation curve of Q_i and Q_o with cam angle under different c_i/c_o value. When the c_i/c_o value increases, as a result to the influence of oil film pressure, Q_i increases, while Q_o remains basically unchanged. Figure 7.14e, f present the variation curve of μ_i and μ_o with cam angle under different c_i/c_o value. Figure 7.14e presents the variation curve of the speed ratio with cam angle under different c_i/c_o value. First of all, it can be discovered that μ_i increases (T_i decreases) with the increase of the c_i/c_o value. Due to the increase of the c_i/c_o value, Q_i increases, causing a decrease of T_i and an increase of μ_i. And it can also be seen that the speed ratio decreases with the increase of the c_i/c_o value. This is because that the viscous resistance torque increases on the inner side of the floating bush, resulting in the speed of the floating bush decreases and the bush speed ratio decreases. In addition, it can be seen that when the c_i/c_o value increases, T_o increases. The reason is that the decrease in floating bush speed enlarges the relative speed of the outer layer which increases the heat generation of the outer layer.

7.5.4 Dynamic Characteristic Analysis

7.5.4.1 The Influence of Operating Conditions Parameters

(1) The Influence of Cam Speed

Figure 7.15 shows the relative motion trajectories of the floating bush and roller as the cam speed varies. Figure 7.15a, b show the relative motion trajectory of the floating

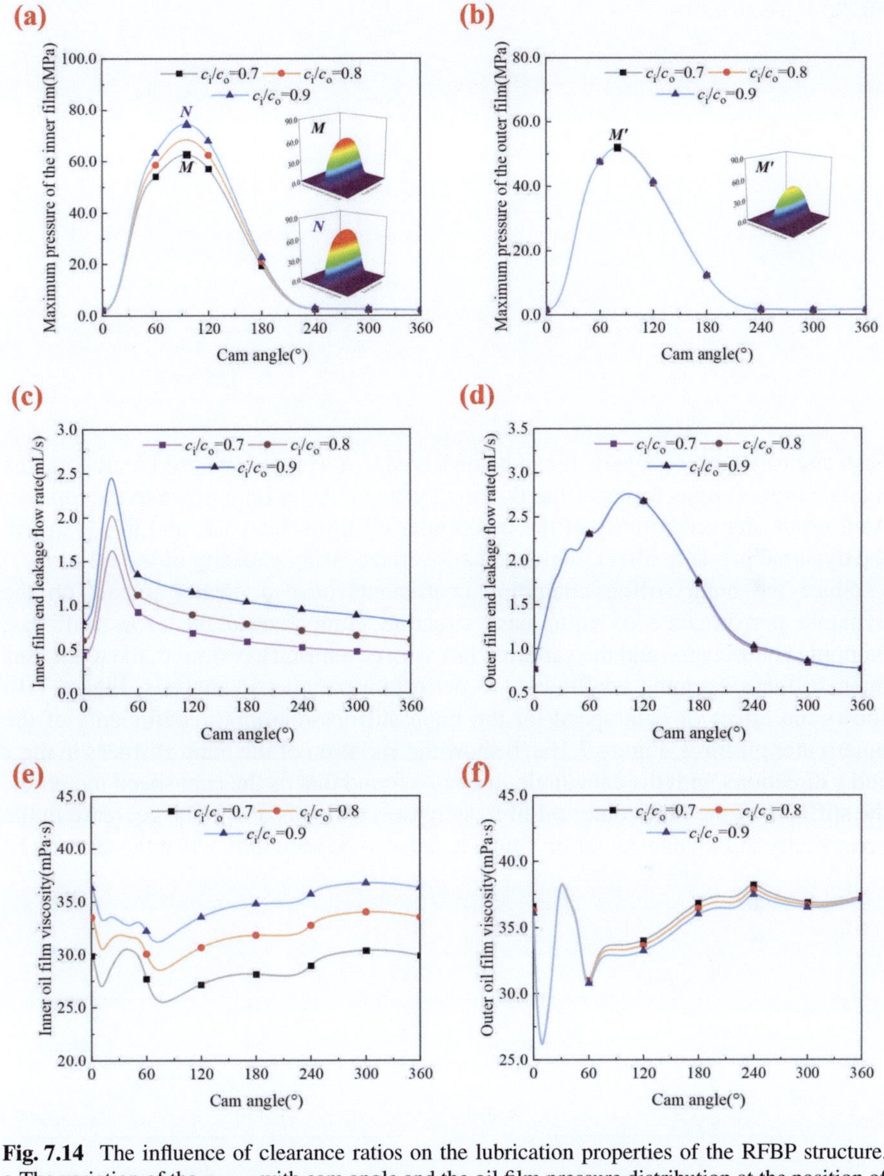

Fig. 7.14 The influence of clearance ratios on the lubrication properties of the RFBP structure. **a** The variation of the $p_{\mathrm{i,max}}$ with cam angle and the oil film pressure distribution at the position of the maximum value of $p_{\mathrm{i,max}}$; **b** The variation of the $p_{\mathrm{o,max}}$ with cam angle and the oil film pressure distribution at the position of the maximum value of $p_{\mathrm{o,max}}$; **c** The variation of Q_{i} with cam angle; **d** The variation of Q_{o} with cam angle; **e** The variation of μ_{i} with cam angle; **f** The variation of μ_{o} with cam angle; **g** The variation of speed ratio (ω_r/ω_2) with cam angle

Fig. 7.14 (continued)

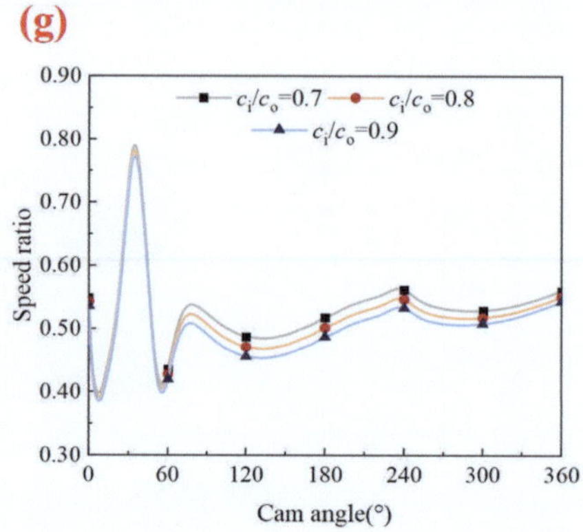

bush and roller, respectively. It can be observed that as the cam speed increases, the relative motion trajectories of the floating bush and roller both move to the top left. As a result, the eccentricity of the inner/outer oil films decrease, and this is due to the dynamic pressure effect, which enhances the bearing capacity of the oil films.

Since the main stiffness/damping coefficients have a greater impact on the dynamic performance of equipment structure compared to the cross stiffness/damping coefficients, and the variation law with eccentricity is similar, the work uses main stiffness/damping coefficients to perform a parametric analysis. Figure 7.16 shows the effect of cam speed on the main stiffness/damping coefficients of the inner/outer oil films. Figure 7.16a, b show the variation of the main stiffness in the x and y directions with the cam angle. It can be found that as the cam speed increases, the stiffness of the inner/outer oil film decreases, which is due to the decrease in the eccentricity of the inner/outer oil films. It can also be seen that, when the cam speed

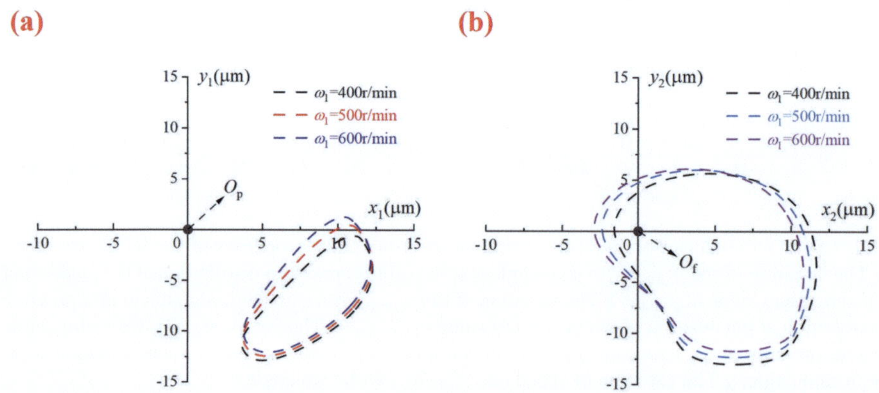

Fig. 7.15 The influence of cam speed on the relative motion trajectories of floating bush and roller (Cam speed is set to 400, 500 and 600 r/min). **a** Relative motion trajectory of floating bush; **b** Relative motion trajectory of roller

increases, the stiffness of the inner oil film decreases more significantly than the stiffness of the outer oil film. The reason is that the inner oil film structure is smaller, the change in cam speed has a greater impact on the lubrication characteristics of the inner oil film. Figure 7.16c, d show the variation of the main damping in the *x* and *y* directions with the cam angle. It can be found that, similar to the stiffness of the oil film, when the cam speed increases, the damping of the inner/outer oil films decreases, and the damping of inner oil film changes more significantly compared to the damping of the outer oil film.

Figure 7.17 shows the effect of cam speed on the equivalent main stiffness/damping coefficients of RFBP structure. Figure 7.17a, b show the equivalent main stiffness and damping coefficients, respectively. It can be seen that as the cam speed increases, both the equivalent main stiffness coefficients and the equivalent main

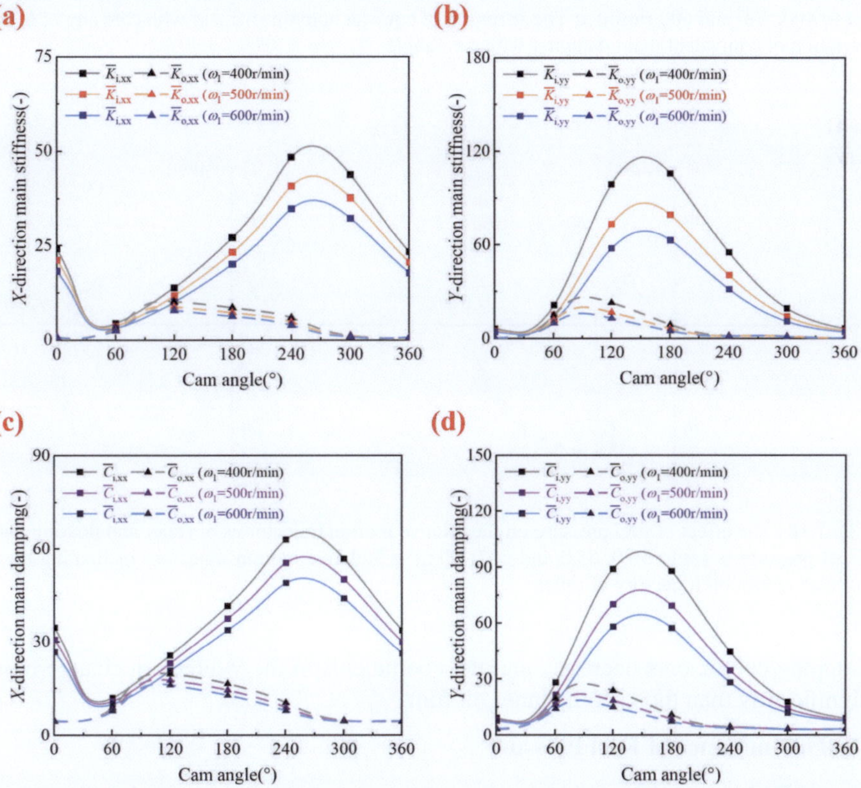

Fig. 7.16 The effect of cam speed on the main stiffness/damping coefficients of the inner/outer oil film (Cam speed is set to 400, 500 and 600 r/min). **a** The variation of *x*-direction main stiffness with cam angle; **b** The variation of *y*-direction main stiffness with cam angle; **c** The variation of *x*-direction main damping with cam angle; **d** The variation of *y*-direction main damping with cam angle

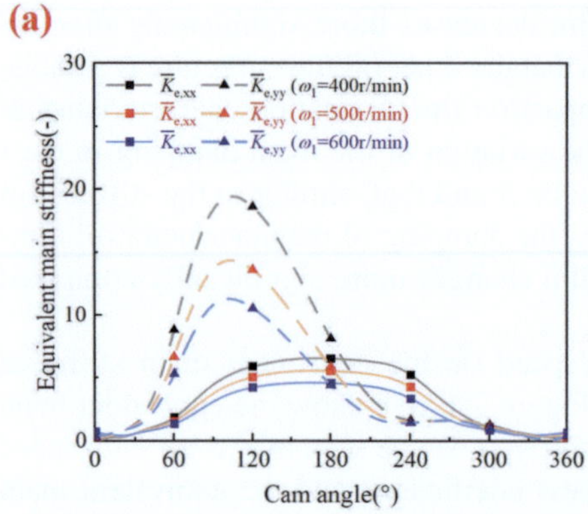

Fig. 7.17 The effect of cam speed on the equivalent stiffness/damping coefficients (Cam speed is set to 400, 500 and 600 r/min). **a** The variation of equivalent main stiffness with cam angle; **b** The variation of equivalent main damping with cam angle

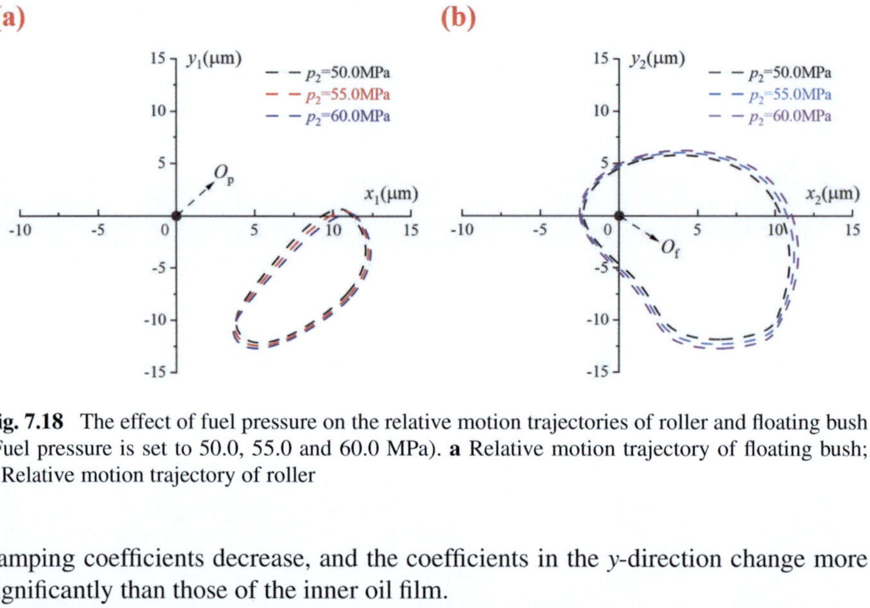

Fig. 7.18 The effect of fuel pressure on the relative motion trajectories of roller and floating bush (Fuel pressure is set to 50.0, 55.0 and 60.0 MPa). **a** Relative motion trajectory of floating bush; **b** Relative motion trajectory of roller

damping coefficients decrease, and the coefficients in the *y*-direction change more significantly than those of the inner oil film.

(2) The Influence of Fuel Pressure

Figure 7.18 shows the relative motion trajectories of the floating bush and roller as the fuel pressure varies. Figure 7.18a, b show the relative motion trajectory of the floating bush and roller, respectively. The work adjusts the maximum fuel pressure (p_2) to change the fuel pressure. It can be observed that the relative motion trajectories of the floating bush and roller both tend to move to the lower right, which results in an increase in the eccentricity of the inner/outer oil film. The reason is that as the fuel pressure increases, the external load on the RFBP structure increases.

Figure 7.19 shows the effect of fuel pressure on the main stiffness/damping coefficients of the inner/outer oil films. Figure 7.19a, b show the variation of the main stiffness in the x and y directions with the cam angle. It can be found that as the fuel pressure increases, the stiffness of the inner/outer oil film increase, as a result of the oil films eccentricity increase. Moreover, when the fuel pressure increases, the stiffness of the inner oil film increases more significantly. Figure 7.19c, d show the variation of the main damping in the x and y directions with the cam angle. It can be found that, similar to the stiffness of the oil film, when the fuel pressure increases, the damping of the inner/outer oil films decreases. And when the fuel pressure increases, the inner oil film damping changes more significantly.

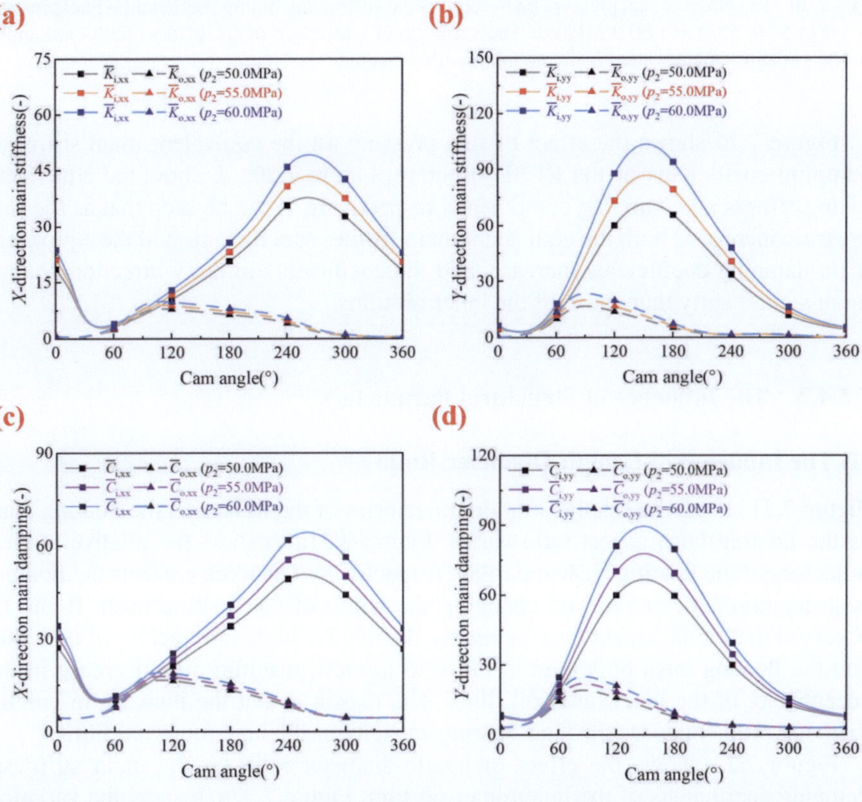

Fig. 7.19 The effect of fuel pressure on the main stiffness/damping coefficients of inner/outer oil film (Fuel pressure is set to 50.0, 55.0 and 60.0 MPa). **a** The variation of x-direction main stiffness with cam angle; **b** The variation of y-direction main stiffness with cam angle; **c** The variation of x-direction main damping with cam angle; **d** The variation of y-direction main damping with cam angle

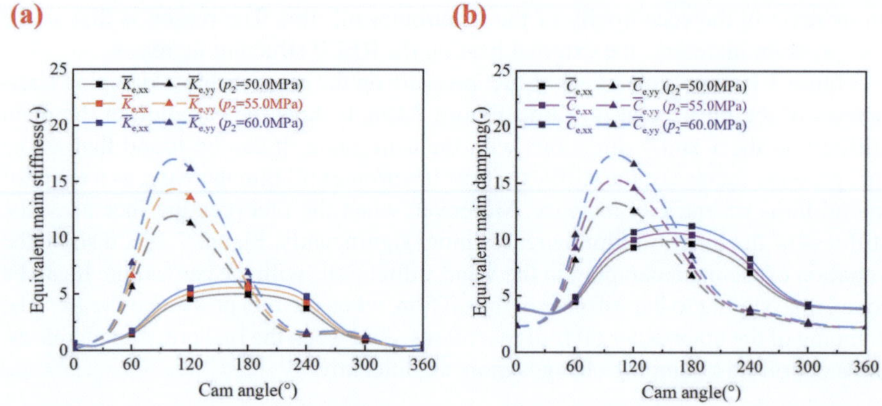

Fig. 7.20 The effect of fuel pressure on the equivalent stiffness/damping coefficients (Fuel pressure is set to 50.0, 55.0 and 60.0 MPa). **a** The variation of equivalent main stiffness with cam angle; **b** The variation of equivalent main damping with cam angle

Figure 7.20 shows the effect of fuel pressure on the equivalent main stiffness/damping coefficients of the RFBP structure. Figure 7.20a, b show the equivalent main stiffness and damping coefficients, respectively. It can be seen that as the fuel pressure increases, both the equivalent main stiffness coefficients and the equivalent main damping coefficients increase, and the coefficients in the y-direction change more significantly than those of the inner oil film.

7.5.4.2 The Influence of Structural Parameters

(1) The Influence of Length-Diameter Ratio

Figure 7.21 shows the relative motion trajectories of the roller and the floating bush as the floating bush aspect ratio varies. Figure 7.21a, b show the relative motion trajectory of the floating bush and roller, respectively. The work controls the floating bush length-diameter ratio by changing the width of the floating bush. It can be observed that as the aspect ratio increases, the relative motion trajectory of the roller and the floating bush both tend to move to top left, resulting in a decrease in the eccentricity of the inner/outer oil films. The reason is that the increase in length-diameter ratio enhances the load-bearing capacity of the inner/outer oil films.

Figure 7.22 shows the effect of length-diameter ratio on the main stiffness/damping coefficients of the inner/outer oil film. Figure 7.22a, b show the variation of the main stiffness in the x and y directions with the cam angle. It can be found that as the aspect ratio increases, the stiffness of the inner/outer oil film decreases, as a result of the oil film eccentricity decreases. Moreover, when the length-diameter ratio increases, the stiffness of the inner oil film decreases more significantly. Figure 7.22c, d show the variation of the main damping in the x and y directions with the cam angle.

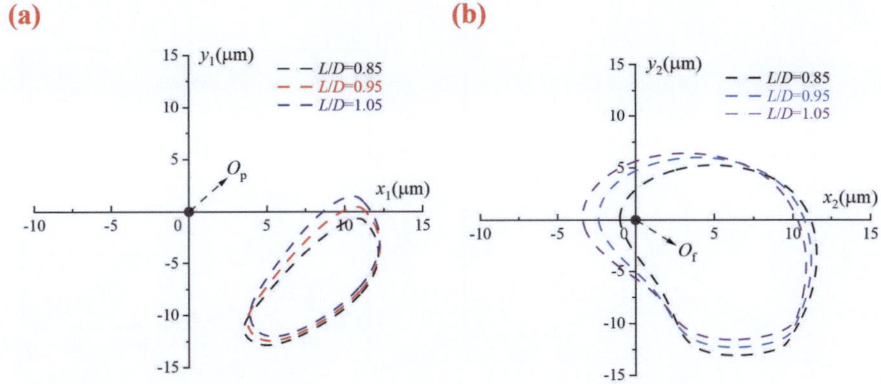

Fig. 7.21 The effect of length-diameter ratio on the relative motion trajectories of roller and floating bush (Length-diameter ratio is set to 0.85, 0.95 and 1.05). **a** Relative motion trajectory of floating bush; **b** Relative motion trajectory of roller

It can be found that, similar to the oil film stiffness, when the aspect ratio increases, the inner/outer oil film damping decreases, and the inner oil film damping changes more significantly.

Figure 7.23 shows the effect of aspect ratio on the equivalent main stiffness/damping coefficients of RFBP structure. Figure 7.23a, b show the equivalent main stiffness and damping coefficients, respectively. It can be seen that as the length-diameter ratio increases, the equivalent main stiffness/damping coefficients decrease, and the *y*-direction equivalent main stiffness/damping coefficients change more significantly.

(2) The Influence Clearance Ratio

Figure 7.24 shows the relative motion trajectories of the floating bush and roller as the clearance ratio (c_i/c_o) varies. Figure 7.24a, b show the relative motion trajectory of the floating bush and roller, respectively. The clearance ratio is adjusted by fixing the outer clearance and changing the inner clearance. It can be observed that as the clearance ratio increases, the relative motion trajectory moves downward to the left, as a result, the eccentricity of the inner oil film increases. The reason is that the inner oil film bearing capacity decreases which is caused by the higher clearance ratio. And due to the inner clearance increases, the range of motion of the floating bush increases. In addition, because the outer clearance remains unchanged, the load-bearing capacity of the outer oil film and the relative motion trajectory of the rollers also remain essentially unchanged.

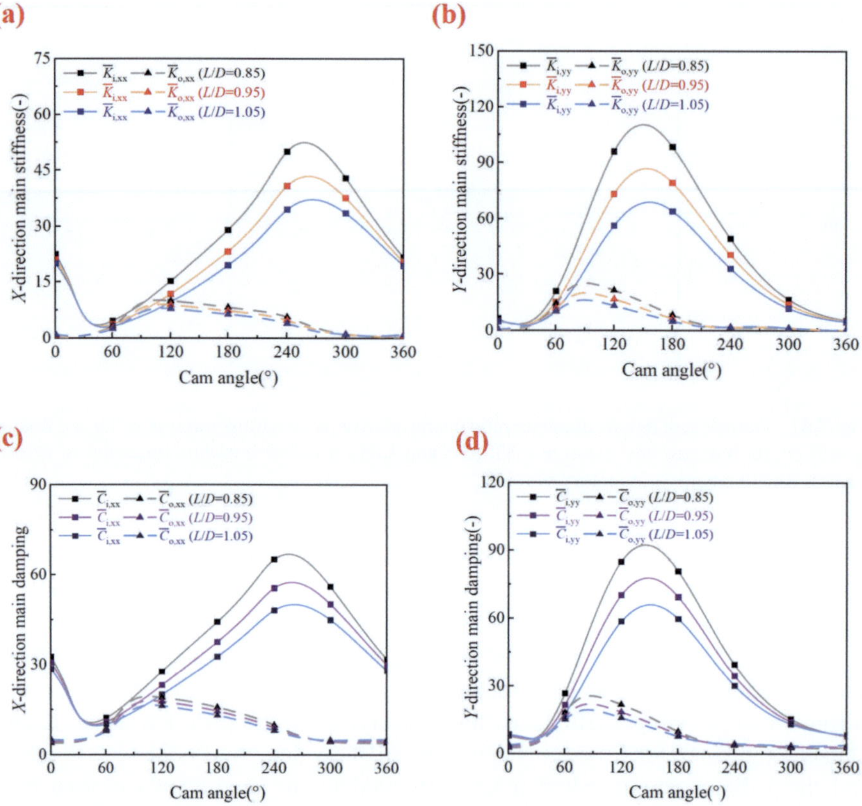

Fig. 7.22 The effect of aspect ratio on the main stiffness/damping coefficients of the inner/outer oil film (Length-diameter ratio is set to 0.85, 0.95 and 1.05). **a** The variation of *x*-direction main stiffness with cam angle; **b** The variation of *y*-direction main stiffness with cam angle; **c** The variation of *x*-direction main damping with cam angle; **d** The variation of *y*-direction main damping with cam angle

Figure 7.25 shows the effect of clearance ratio on the main stiffness/damping coefficients of the inner/outer oil film. Figure 7.25a, b show the variation of the main stiffness in the x and y directions with the cam angle. It can be found that as the clearance ratio increases, the stiffness of the inner oil film increases, and the stiffness of the outer oil film remains basically unchanged. The reason for this is that as the clearance ratio increases, the load-bearing capacity of the inner layer becomes weak and the outer layer remains basically unchanged. Figure 7.25c, d show the variation of the main damping in the x and y directions with the cam angle. It can be found that, similar to the oil film stiffness, when the clearance ratio increases, the inner oil film damping increases, and the outer oil film damping remains basically unchanged.

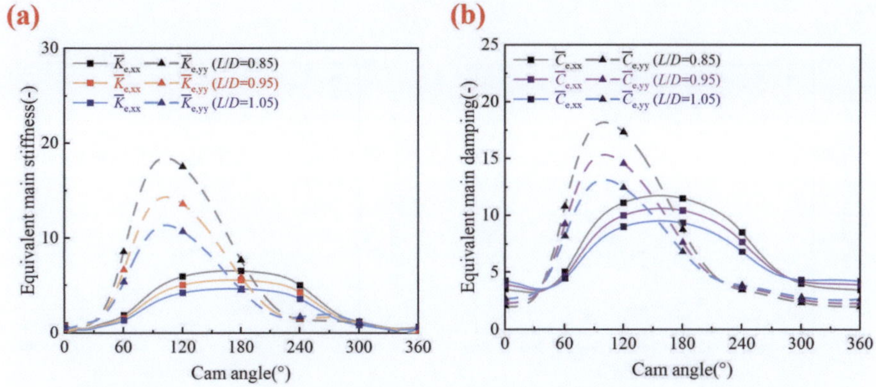

Fig. 7.23 The effect of aspect ratio on the equivalent stiffness/damping coefficients (Length-diameter ratio is set to 0.85, 0.95 and 1.05). **a** The variation of equivalent main stiffness with cam angle; **b** The variation of equivalent main damping with cam angle

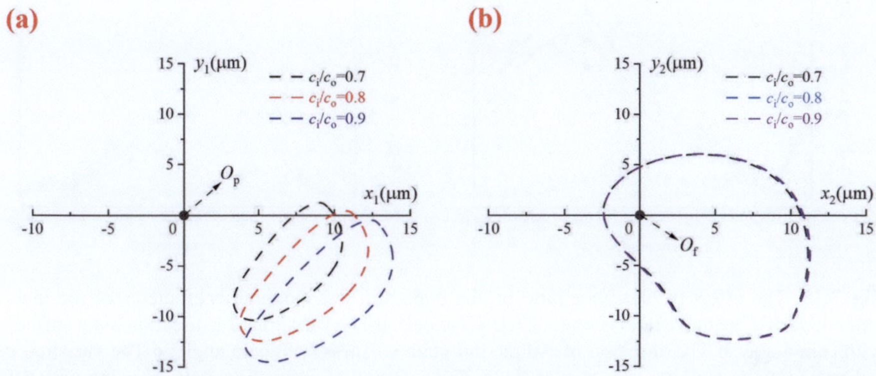

Fig. 7.24 The effect of clearance ratio on the relative motion trajectories of roller and floating bush (Clearance ratio is set to 0.7, 0.8 and 0.9). **a** Relative motion trajectory of floating bush; **b** Relative motion trajectory of roller

Figure 7.26 shows the effect of clearance ratio on the equivalent main stiffness/ damping coefficients of RFBP structure. Figure 7.26a, b show the equivalent main stiffness and damping coefficients, respectively. It can be seen that as the clearance ratio increases, the equivalent main stiffness/ damping coefficients increase, and the y-direction equivalent main stiffness/damping coefficients change more significantly.

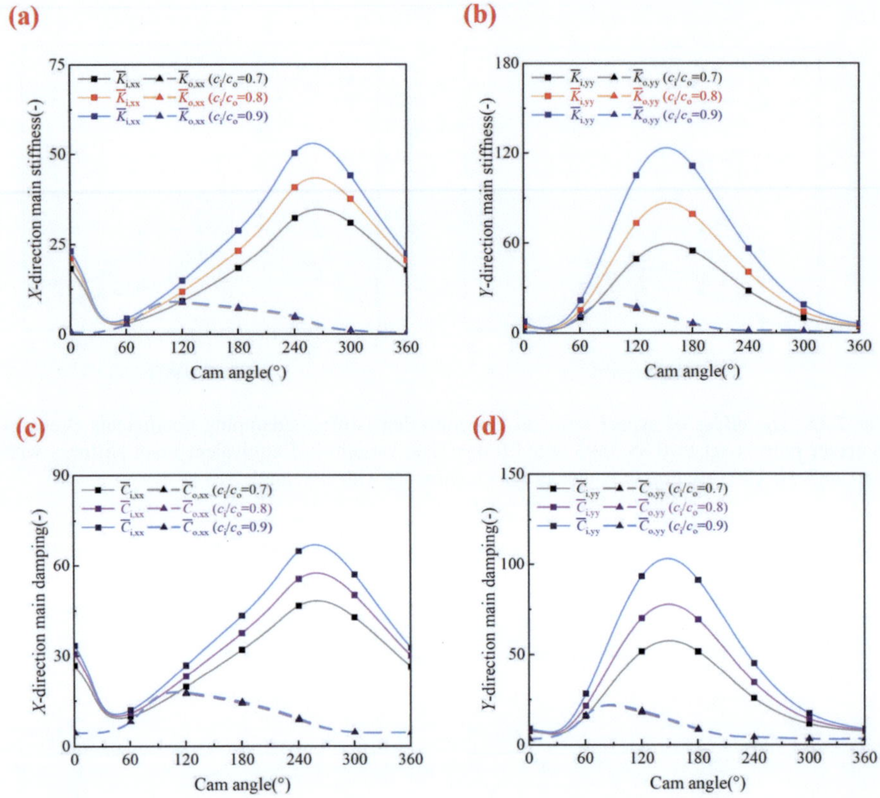

Fig. 7.25 The effect of clearance ratio on the main stiffness/damping coefficients of the inner/outer oil film (Clearance ratio is set to 0.7, 0.8 and 0.9). **a** The variation of x-direction main stiffness with cam angle; **b** The variation of y-direction main stiffness with cam angle; **c** The variation of x-direction main damping with cam angle; **d** The variation of y-direction main damping with cam angle

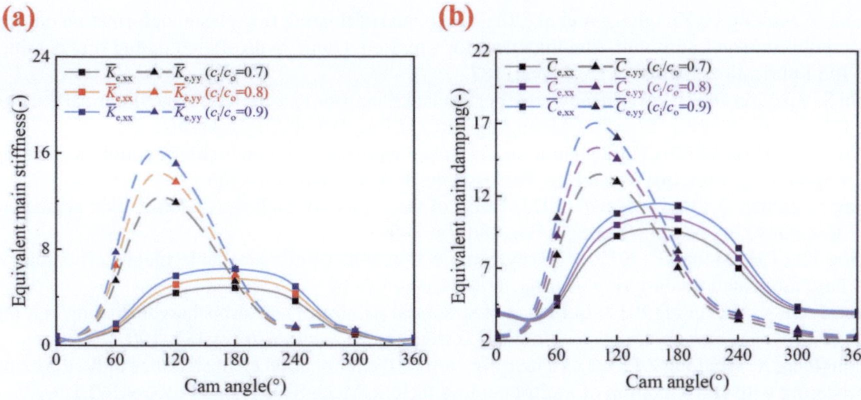

Fig. 7.26 The effect of clearance ratio on the equivalent stiffness/damping coefficients (Clearance ratio is set to 0.7, 0.8 and 0.9). **a** The variation of equivalent main stiffness with cam angle; **b** The variation of equivalent main damping with cam angle

References

Alakhramsing SS, Rooij DBM, Schipper JD et al (2018) A full numerical solution to the coupled cam-roller and roller-pin contact in heavily loaded cam-roller follower mechanisms. Proc Inst Mech Eng, Part J: J Eng Tribol 232(10):1273–1284

Andrés SL, Kerth J (2004) Thermal effects on the performance of floating ring bearings for turbochargers. Proc Inst Mech Eng, Part J: J Eng Tribol 218(5):437–450

Andres SL, Rivadeneira CJ et al (2007) Rotordynamics of small turbochargers supported on floating ring bearings-highlights in bearing analysis and experimental validation. J Tribol 129(2):391–397

Bin G, Huang Y, Guo S et al (2018) Investigation of induced unbalance magnitude on dynamic characteristics of high-speed turbocharger with floating ring bearings. Chinese J Mech Eng 31(1):88

Chao Q, Yang K, Kun S et al (2017) Effect of surface roughness on static characteristics of floating ring bearing. Lubr Eng 42(5):66–72

El-Ghoul Z, John S, Dimkovski Z et al (2019) Wear analysis of pin and roller surfaces. J Phys: Conf Ser 1183(1):012007

Holt C, San Andre SL, Sahay S et al (2005) Test response and nonlinear analysis of a turbocharger supported on floating ring bearings. J Vib Acoust 127(2):107–115

Hong G, Boqian X, Shaoqi C (2011) Performance analysis of high speed floating ring hybrid bearing in the laminar and turbulent regimes. Adv Mater Res 1167(197–198):1776–1780

Hong G, Xinmin L, Xiaoling W et al (2009) Performance of flat capillary compensated deep/shallow pockets hydrostatic/hydrodynamic journal-thrust floating ring bearing. Tribol Trans 52(2):204–212

Liqiang P, Lixin W, Zhiying Q et al (2023) Oil-film forming behavior of floating ring bearing under oil-starved lubrication. Wear 524–525

Novotný P, Skara P, Hliník J (2018) The effective computational model of the hydrodynamics journal floating ring bearing for simulations of long transient regimes of turbocharger rotor dynamics. Int J Mech Sci 148:611–619

Pei S, Xu H, Yun M et al (2016) Effects of surface texture on the lubrication performance of the floating ring bearing. Tribol Int 102(1):143–153

Shuai Y, Junxing H, Xiaodong A et al (2023) Influence of floating ring elastic deformation on ring equilibrium and minimum film thickness for a hydrodynamic/hydrostatic floating ring bearing. Ind Lubrication Tribol 75(10):1178–1187

Soni S, Vakharia D (2017a) A steady-state performance analysis of a non-circular cylindrical floating ring journal bearing. Proc Inst Mech Eng Part J: J Eng Tribol 231(1):41–56

Soni S, Vakharia D (2017b) Dynamic performance analysis of a noncircular cylindrical floating ring bearing. Proc Inst Mech Eng, Part J: J Eng Tribol 231(6):745–765

Yang K, Zizhen Q, Qiming F et al (2023) Effect of the number of oil holes on lubrication characters of floating ring bearing. Lubricat Eng 48(09):83–90

Wang X, Li H, Meng G (2017) Rotordynamic coefficients of a controllable magnetorheological fluid lubricated floating ring bearing. Tribol Int 114:1–14

Zhao J, Gu Y, Yi S et al (2017) Influence of structural parameters of turbocharger floating bearing on its dynamic characteristic coefficients. J Beijing Inst Technol 26(2):183–190

Zhongliang X, Weidong Z (2022) An investigation on the lubrication characteristics of floating ring bearing with consideration of multi-coupling factors. Mech Syst Signal Process 162:108086

MIX
Papier aus verantwortungsvollen Quellen
Paper from responsible sources
FSC® C105338

If you have any concerns about our products,
you can contact us on
ProductSafety@springernature.com

In case Publisher is established outside the EU,
the EU authorized representative is:
Springer Nature Customer Service Center GmbH
Europaplatz 3, 69115 Heidelberg, Germany

Printed by Libri Plureos GmbH
in Hamburg, Germany